IN-CAR ENTERTAINMENT MANUAL

Choosing □ Installing □ Improving

Dave Pollard

First edition published 1989
Reprinted 1991
Revised Second Edition
published 1993
Reprinted 1995 and 1996

© Dave Pollard 1993

All rights reserved. No part of this book may be reproduced or transmitted in any form or by any means, electronic or mechanical, including photocopying, recording or by an information storage or retrieval system, without permission of the publisher.

Published by:
Haynes Publishing,
Sparkford, Nr Yeovil, Somerset
BA22 7JJ, England

British Library Cataloguing in Publication Data

A catalogue record for this book is available from the British Library.

ISBN 1 85010 862 5

Printed in Great Britain by
J. H. Haynes & Co. Ltd.

While every effort is taken to ensure the accuracy of the information given in this book, no liability can be accepted by the author or publishers for any loss, damage or injury caused by errors in, or omissions from, the information given.

Contents

	Page
Acknowledgements	4
Introduction	
Planning, preparation and building a system	5
Chapter 1	
Tools, techniques and workshop procedures	13
Chapter 2	
Connections, wire and fuses	27
Chapter 3	
Speakers and crossovers, including sub-woofers and sub-bass tubes	34
Chapter 4	
Cassettes, description, care and storage	68
Chapter 5	
Choosing and fitting a radio/cassette deck	77
Chapter 6	
Choosing and fitting an aerial	97
Chapter 7	
Dealing with interference	103
Chapter 8	
Compact discs, description, care and storage	109
Chapter 9	
Choosing and fitting a CD tuner and autochanger	114
Chapter 10	
Choosing and fitting a graphic equaliser and DSP	132
Chapter 11	
Choosing and fitting a power amplifier	140
Chapter 12	
In-car telephones	150
Chapter 13	
Security	164
Appendix	
Useful names and addresses	177
Glossary of terms	183

Acknowledgements

Every manual you are ever likely to read is merely the tip of a very large literary iceberg and this one is no exception. Therefore, I have to proffer my grateful thanks to all of the following who provided assistance in the form of expertise and guidance and also in providing equipment for photography purposes;

Graham Johnson (Alpine Electronics of UK Ltd.)
Neil Braybrook (Braybrook Car Radio)
Christopher Parsons (Clarion Shoji (UK) Ltd)
Paul Dards (Dards Electronics Ltd)
Tim Snelguard (Kamasa Tools Ltd)
Paul van der Marck (Philips Car Stereo)
Christine Whitworth (Sykes-Pickavant Ltd)
Colin North (Teng Tools Ltd)
Richard Tyminski (Truck King)
Tim Elsey (Panasonic Car Phones)
Trevor Preece & Kevin O'Byrne (Car Stereo & Security magazine)
Philippe Plateau (Fischer C-Box)

FOR THE RECORD

George McLeod of Alpine Electronics of UK Ltd., was an absolute mine of technical information and was very helpful during the preparation of this book; many technical 'I's and 'T's have been dotted and crossed with his assistance. Paul Dards (Dards Electronics, Milton Keynes) played a major role in lending his premises, cars and skilled staff (thanks Tim and Barry) for several installation sections.

My wife Ann has, of course, played her part, checking and re-checking copy and photographic references.

Roy Craggs is a personal friend who helped out on several other projects before being coerced into this one. I have to thank him sincerely, not only for his help on this one, but also in anticipation of his help on the next!

If you've smiled at the superb illustrations and caricatures included in this edition, then you'll have become, as I have, a fan of Nicky Smart. Through his magic pen he has truly captured those mystical elements that come together to form the motor car enthusiast. Line diagrams not provided by the manufacturers (and acknowledged as such) were prepared by Davan Designs.

DEDICATION

To George and Margaret who made everything possible and to Phil and Nita for putting me on the Right Track. Most of all to Ann for her constant support and for believing when no one else did.

Introduction

The *In-car Entertainment Manual* aims to cover every aspect of in-car entertainment from a practical point of view and one that can be easily understood. There is a brief description of the principles involved (for example, the nature of radio signals), kept deliberately simple and free from jargon. It is doubtful if the book will make you a hi-fi wizard (there's not enough space for that) but it should give you a greater insight into the massively expanding area of modern motoring; not only the knowledge of what you are buying, but also to know what it is you require **before** you buy it!

Throughout the book, the acronym ICE is used to denote the phrase In-Car Entertainment. Audio systems in cars aren't new, though there has been a massive increase in its popularity, which started in the 1970s and has snowballed dramatically since. The first car radios were not particularly popular, as they were large, usually needed a transformer and, as the cars themselves were particularly noisy places to be, a car radio was pretty much a waste of time. Little wonder, then, that the Americans, with their predilection for large (and quiet) engined limousines got a head start in ICE. Improvements in car and engine design meant that cars have become quicker, cheaper and quieter just as the transistor did away with the need for large, cumbersome valves. Yet another new age dawned when the advent of micro-chip technology enabled incredibly complex electronic circuits to be placed literally in the palm of your hand.

With the driver now able to hear himself think (and becoming quite bored in the process), this miniaturisation was an obvious boon and when Philips made the Compact Cassette, reliable in-car tape players 'took off' in a big way.

Not all ICE has been so successful, though. In the early 1970s, the eight-track cartridge was hailed as a viable alternative to the cassette. By the end of the decade, a car with an eight-track deck in situ was rare indeed. This one example proves a very important point; in order to be a feasible in-car proposition, any piece of in-car audio equipment/software needs also to be popular on the domestic market. The home cassette player caught-on in a big way, the eight-track didn't. The past few years have seen yet another Philips innovation – the Compact Disc. Here, we can again apply the 'popular at home' principle. The Compact Disc (CD) caught the imagination of the world at large and with lots of manufacturers making the equipment (both domestic and in-car) the cost of purchasing a player has dropped dramatically over the relatively few years since its inception.

World technology moves at an ever increasing rate and few areas can keep up with the pace set by in-car electronics. When the first edition of this book came out, the buzzword (or buzz-abbreviation) was DAT - Digital Audio Tape.

DAT was, according to many,

Introduction: planning, preparation and building a system:

going to revolutionise the world of ICE, take up the mantle of the compact cassette and defeat the compact disc. But little is ever what it seems. In the first place, the compact cassette was not quite ready to roll over and die, with the result that cassette sales have remained steady. Secondly, compact disc sales (and those of domestic and in-car machinery) have absolutely rocketed, to the point where CDs now heavily outsell their immediate rivals, the vinyl disc. The failure of DAT to make any significant impact on the audio scene can be linked to; a) the cost of the equipment b) a distinct lack of domestic equipment and c) the cost and lack of choice of prerecorded tapes. This latter point was aggravated by the fact that the major protagonists in the DAT machinery market could not agree terms with regard to royalties and the law appertaining to copyright etc. The abilities of DAT are unquestioned and within the music industry (recording studios etc.), it is highly regarded and well used. In terms of the overall marketplace, however, it is a drop in the ocean. Stranger things have happened at sea, but it now seems highly unlikely that DAT will ever come to the prominence once predicted for it.

The Compact disc, however, has gone from strength to strength. The massive growth in the domestic market made it necessary for there to be a wide selection of discs for the users to choose from. As such, this fuelled an increase in in-car equipment. Higher sales of both domestic and in-car units meant that more manufacturers joined in and the result of this was that sets became ever more impressive in terms of sound and features, whilst becoming ever cheaper.

On the horizon is Philips DCC – Digital Compact Cassette. It is tempting to ask the question – why should Philips succeed with DCC where others failed (by enlarge) with DAT? Well, given that Phillips were responsible for the original Compact Cassette and the Compact Disc, it would be a foolish man who would dismiss any ideas from this company out of hand; and then there's the technical specification.

Like DAT, DCC offers CD-like quality from tape. Unlike DAT, DCC is backward compatible, meaning that standard cassettes can be used on a DCC deck. Clearly, this opens up the market considerably. It is fair to prognosticate that the CD market will continue to boom at the expense of the vinyl disc and that the cassette market will continue to boom, with the sales split between the original and the digital versions.

The many technical terms used to describe the features of ICE equipment are understood by precious few purchasers, despite the fact that they are shelling out their hard-earned cash! Would you buy a car without knowing the size of the engine or how many doors it had? I think not. The ability to tell how good (or otherwise) a piece of equipment is by reading the side of the box is something the reader should gain by reading this book. Reading the dedicated ICE magazines is a way to discover what the experts think of various pieces of equipment – especially useful if there is a 'comparison' test. At the time of writing, there are two such publications on the shelves, so there's no lack of choice.

Chapters one and two are preparatory, but vitally important. Chapter one is a brief listing of the tools that make life much easier when installing your ICE equipment. Chapter two gives a brief description of the most commonly used connections and how to fit them. It also includes details of fuses. Wherever possible, pictures and/or diagrams have been used to make descriptions clearer.

In-car audio is an area riddled with so much jargon, it's easy to believe that you are dealing with another language. As such, a glossary of the most common terms has been included at the end of this book. Ideally, it would cover every single term you are ever likely to come across in your search for hi-fi excellence. There are two problems here; firstly, most manufacturers like to have their own designations for various functions and secondly, new terms are being introduced at such a rate that it would be impossible to keep up. It does not include specific technical terms which would only be of use to qualified audio engineers. As with most areas of DIY, having sufficient time and patience is a must. Although you may be itching to get out there and fit your new radio/cassette deck, half an hour spent in your living room, flicking through the instructions (and this book, of course!) would be time well spent. Indeed, it could save a lot of time, money and wear and tear on your nerves.

PREPARATION

Whether or not you studied Baden-Powell's doctrines, it certainly makes sense to 'be prepared' when fitting ICE to your car. Preparation begins in the store where you buy your equipment (assuming you buy it new). Take some time to make sure that you are selecting what you really want, especially when buying a front-end unit, such as a radio/cassette deck. This is covered in more detail in later chapters. Don't discount the

Introduction: planning, preparation and building a system:

specialist dealers, either. Ten years ago, they were definitely the place to go if you were an enthusiast with a deep pocket. But nowadays, there are more of them and, often, their prices can be on a par with the high-street discount merchants. Moreover, they can usually offer advice about equipment and its fitting which cannot be obtained from a 'general' accessory store, where staff are expected to be real 'Jack of all trades'.

When you make your choice, open the box and look inside to see what is supplied with it. With a front-end unit, you need to know is there is anything else you will need to buy in order to fit the set. The retailer should be happy to let you do this if you are going to buy the set, especially if you end up buying a few extras! Some manufacturers go to a lot of trouble and include everything but the kitchen sink, with adaptors and connectors for all eventualities. This means you'll lots of bits left over to pop in your 'odd bits and pieces' box, that magical relative of Pandora which will always yield that elusive nut, screw or bolt you need in a hurry. Throw them away, and you can guarantee you'll need them almost immediately!

Some manufacturers include only the bare minimum and you made need to purchase a little bracketry and a connector or two in order to get going. If you don't do this now, there's the temptation to 'bodge' the set when you're out in the garage. Also, check what fittings are on the ends of the power and speaker leads. These areas are by no means standardised and there are fairly good odds that the leads already in your car (if any, check first) will not match those on the new set. It is always handy to have a few Scotchlok connectors in your box and make sure that you have enough wire for the set.

Whilst using up odd bits of wire you have lying around in the garage may be economical, you'll end up using all sorts of different colours (or worse, all the same colour!) and possibly different thicknesses.

Doing this is potentially quite dangerous and, in the event of the set not working, could drive you out of your mind trying to sort out what goes where. Don't forget that speakers need to be connected using wire designed specially for speakers, it's called speaker wire. Don't strip out a lead from a piece of mains cable and use it for speakers! Just as important is that you don't use speaker cable for power applications. Having said this, most makers supply a generous amount of cable with their speakers, but it's always worth having some in your garage, just in case. Having accumulated the necessary parts, take some time to plan the task. Always have enough time to do it right. Fitting audio equipment can be tricky and time-consuming and things don't always go according to plan. Working to a tight schedule almost guarantees failure and usually a little raised blood pressure as well! Take a tip from the professionals; when fitting a complete audio system (i.e. head unit and speakers) start by fitting the speakers first and work backwards towards the set.

Last, but by no means least, always make sure that you have a few spare fuses in your tool box before the off. One spare just isn't enough, as even a simple fault could blow two or three before you find the cause of your grief. As most of use don't get around to fitting our 'new toys' until Sunday morning, these simple preparations can save an awful lot of heartbreak and frustration, not least for your 'better half' who has to listen to your rantings. There are many motoring stores which open seven days a week, but even they won't be much help if your car is sitting lifeless on the driveway with a system full of 'dead' fuses and an obscure wiring fault!

BUYING SECONDHAND ICE
Buying used ICE equipment is as big a minefield as buying a used car and carries many similar caveats. New sets are not particularly expensive nowadays and you should bear this in mind; could you wait a while and buy new kit? Always make sure that you listen to the equipment before you part with your cash and make sure that the various functions are working (f.fwd, auto-scan etc.). Although most modern ICE gear is very reliable, it is also non-user serviceable, for most of us, at least. As with buying a car, it is always wise to have an 'expert' with you, someone who knows what they should be looking for.

In terms of price, you can take little notice of the price the vendor says he paid new. Whilst he may have paid ú400 for his radio/cassette deck a year ago, you will probably now be able to buy a new set, better specified, for less. If you must buy secondhand, make sure that you make a significant saving over the price now.

BATTERIES
It is vital for the battery to be in good condition in order to have a good audio system. Maintenance is limited to keeping the acid levels topped up with distilled water (unless you have a sealed for life battery) and keeping the terminals free of corrosion. Should corrosion be evident (a white fluffy power on the terminals) the terminals should be removed and cleaned with a wire brush, as should be battery posts. A healthy dab of Copper Ten of similar will ensure that

Introduction: planning, preparation and building a system:

you don't get a repeat performance. Remember that the more you uprate your car's ICE system, the more power will be required from the battery; to a battery already on its last legs, having to cope with the demands of a high powered amplifier may be the last straw.

Even if your battery is brand new, running an uprated system without the engine running could flatten it very quickly.

WHERE TO FIT YOUR ICE

Obvious, but often overlooked; will the equipment you've just bought fit in your car? Cars are made primarily as people and luggage transporters and, to a great extent, ICE equipment (head units and speakers) have to fit around everything else. Nowadays, most manufacturers either offer a deck of some description with the car when new or at least a DIN aperture to fit one of your own choice. This is a good start, as almost every deck you can buy is of DIN specification and will slot straight in. As this aperture is usually in the centre of the car (either at the top of the dash or in the centre console) it is in the best position. The real problems start when you either haven't got a DIN aperture or you have but want to fit more than one DIN item (for example, a radio/cassette deck and a graphic equaliser). Fitting or adapting a centre console to suit your needs is the most obvious answer, though you have to make sure that as well as DIN width and height, you also have DIN depth; some consoles sit directly in front of the car's heating system and a deck could well foul on it.

When it comes to mounting ancillary hardware, such as amplifiers, under the passenger seat is the obvious place to choose. But a sports seat and/or an electric motor in the seat base will drastically reduce the under-seat height available for amplifier fitting. In the boot or hatch is the next place to try - but remember that the petrol tank is usually in that area (unless you have a rear engined car) and it's the simplest thing in the world to drill a hole into it; not to be recommended!

The key to fitting positions is simple; check, measure, think and ask questions before you part with your cash! Tell your dealer what car you have and the system/ equipment you want. He should be able to tell whether it will fit and/or how to fit it. If your car is an older model, you may find that your 'DIN' aperture seems to be too small for your new set - that's because it is ISO DIN. The obvious route is to widen it slightly, but consider the originality of your car - it could remove some of its resale value if you start about it with a hacksaw and a file!

The proof of the pudding; the system installed in this Golf started with a removable radio/cassette deck installed in the dashboard DIN aperture. As more equipment was planned, a VAG centre console was added which had two extra DIN apertures and facilitated.....

... the functional and attractive fitment of a graphic equalizer and a CD player. Note that with the graphic being less than DIN size, a special panel has been made-up to suit. This is not difficult - the easiest way to achieve it being to 'doctor' a standard DIN blanking panel, available either from the manufacturers of the car or from an accessory shop.

BUILDING A SYSTEM

Why bother? A question often levelled by non-ICE enthusiasts, why spend more on your car system than you do on your domestic set-up? Well, if you only use your car for 10 minutes a day, travelling two miles each way to the shops and back, the answer is simple – installing a good system would be a waste of time and money. But many people spend several hours a day in their cars, just travelling to and from work. Of course many actually earn their living on the road and so spend more hours listening to their car stereo in a day than they would in a month of home listening. It follows, therefore, that the place to spend the money is in the car, especially given the largely chaotic state of Britain's roads, where a nice sound system can ease the strain of sitting in a motorway tailback.

If you intend to expand your system at sometime in the future, it is now that you should be planning. For example, if

Introduction: planning, preparation and building a system:

your present radio/cassette deck gives 20W max per channel output, it is tempting just to equip it with speakers capable of handling 30W (see chapter 3, speakers – allow a 50% margin for safety). But if you're going to uprate to 60W per channel next year, then you're not going to have much use out of your present speakers before you have to replace them and buy another set, an expensive business. How much better to buy 90W speakers now, they won't affect the present 20W system (and will probably make it sound better) and when you uprate your power output, you won't have the expense of buying new speakers.

Here are some typical suggested systems, starting at very basic and building to quite complex systems. However, like the cars themselves, the complexity of an in-car hi-fi system knows no limits and for those with a bottomless bank account, you can just keep on adding and modifying as long as you wish.

With any system, the important thing to remember is to consider the future, you really can save a lot of cash by buying good stuff now, rather than later.

Planning your system can save you no small amount of grief – in more ways than one!

The 4-speaker system is becoming very much the industry norm, with ever more vehicle manufacturers taking the hint and fitting such systems as standard in new cars. Indeed, many are now fitting component speakers (mid-range and tweeter) at the front and sometimes at the rear as well.

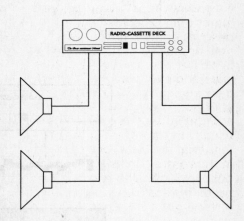

A basic set-up and one very rarely seen today, a 2-speaker, stereo system. However, you should not dismiss such a system out of hand, especially if you are embarrassed in the cashflow department. Rather than stretching yourself to fit a 4-speaker system of indifferent quality, why not buy a good deck and opt for just the front speakers? That way, you'll get great sounds now and when you want to uprate, you can keep the front speakers and add the rear, rather than having to discard four speakers.

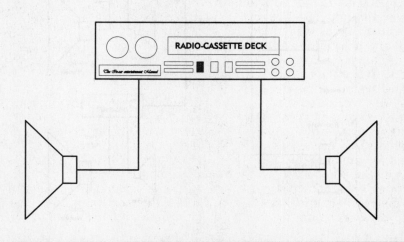

Introduction: planning, preparation and building a system:

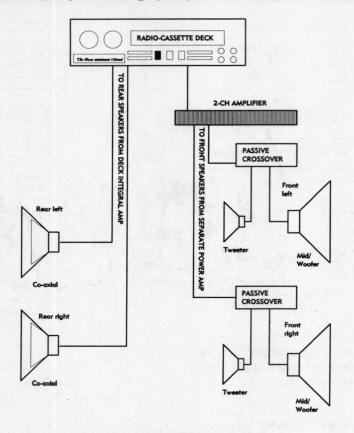

If you deck is suitable for extra amplification, then you can build yourself a really versatile system. This system shows the extra amplifier powering the front speakers, whilst the integral deck amplifier deals with those at the rear. When you add an amp it is usual to go upmarket with the speakers as well. More often than not, the rears are co-axials with more capable component speakers in the front. As you can see, splitting the signal between two speakers requires a crossover, passive, because it comes after the amplifier.

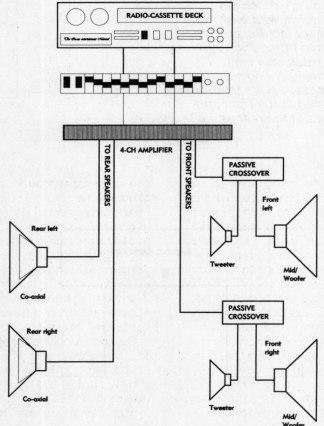

A variant on the last system, this one utilises a 4-channel amplifier and a graphic equaliser as well. In essence, the signals pass from the deck to the equaliser where they can be altered to suit your personal tastes. From there, they go to the amplifier for distribution to the various speakers.

Introduction: planning, preparation and building a system:

Once you start dealing with a pre-amp deck (i.e. one which does not have its own integral amplifier) the sky, or your bank balance, is the limit. Here, the signals are processed by the graphic equaliser which then sorts them out accordingly. Many graphics have a sub-bass facility, which means that the amplifier in question receives only frequencies below (usually) 200 Hz. The rear speakers would be fed by their own amplifier as would the front speakers. Note the addition of a CD autochanger which would make this theoretical system very nice indeed. Whilst it is possible to continue adding and uprating equipment ad nauseum, we are now very definitely at the point where no small amount of skill is involved. In addition, the more power available, the more prone it is going to be to interference problems. In short, from here on in, you need to be a very skilled amateur or an equally skilled professional.

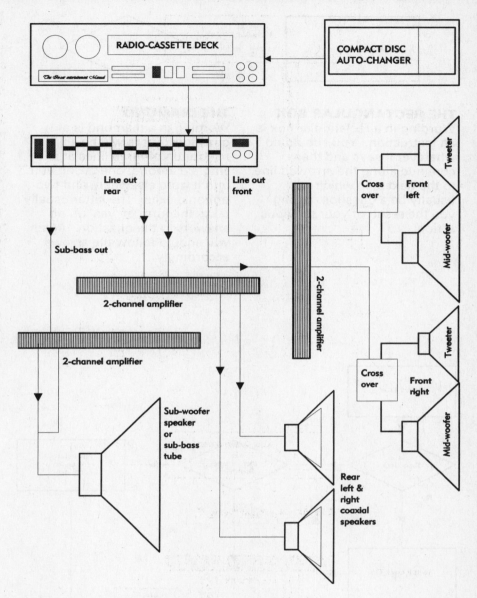

USING A FLOWCHART

On several of the 'fitting' sections, flowcharts are used in order to give a logical sequence of fault finding. Those with experience of computers will be used to them and no explanation will be necessary (skip to the next chapter!). However, for those who are meeting them for the first time, here are a few explanatory notes.

HOW FLOWCHARTS WORK

Although a flowchart can seem, at first glance, a very complex method of solving a fitting problem, quite the opposite is true. In order to keep things as simple as possible, only three symbols are used; a rectangle, a diamond and an ellipse. Once you have mastered the basic techniques, you should be able to apply this method to most areas of in-car DIY.

By creating your own flowchart before attempting a task, you should find that it is of great help when it comes to putting everything back together again. By jotting down a few notes alongside the relevant boxes (how many screws were removed, etc.) life can become much easier.

As mentioned in the 'planning' section, you can save a lot of time and frustration by thinking the job out in advance. As most ICE equipment lights up to show that it is functional, many of the flowcharts start by assuming that you have no lights and, therefore, that the unit is not functioning. You should be aware, however, that on occasion, the fault may simply be that the light source (the bulb or LED) has failed and, in fact, the unit is working perfectly.

Introduction: planning, preparation and building a system:

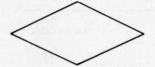

THE RECTANGULAR BOX
Wording in a rectangular box is an instruction. You should do whatever it says and then continue along the arrowed line to the next step which will usually be a question, asking you the result of your previous action.

THE DIAMOND
Wording in a diamond is a question. There will be a maximum of three lines at any single diamond; one taking you into it from elsewhere and two optional exits. The latter usually takes the form of 'yes' or 'no' answers to the question. Again, you should follow the arrows accordingly.

THE ELLIPSE
When you reach an ellipse, you have reached the end of the line. For whatever reason, you will not be able to carry on. There are two ellipse notations. The first is 'Dealer Problem' and can mean that you should ask someone more competent that yourself for their expert opinion or, in some cases, return the unit you have just purchased as being faulty. The second is that you have finished your fault checking procedure and your set is now working perfectly.

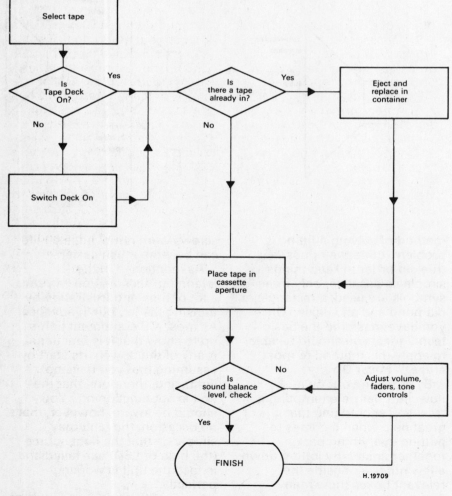

The following is a very simple example of a flowchart application, designed to show its practical usage. It shows how you should tackle the task of putting a cassette tape into your radio/cassette deck. Start at the top and follow the arrows in whichever direction is appropriate. As mentioned earlier, once you understand the logical principles involved, you can soon adapt the flowchart system to your own ends.

Chapter 1
Tools Techniques & Workshop Procedures

If you ever have the opportunity to watched a skilled in-car audio fitter at work*, then take it! As with any professional, you will find it quite amazing that what seemed like complex tasks are so easy. They're not, of course, it just that a professional will make them seem that way. The fitter will use a vast array of highly specialised and expensive tools; and so he should, for his livelihood depends on them. The DIY fitting of ICE, however, does not require such extremes when it comes to tools. To purchase special tools that will only be used once every five years makes no economic sense and negates the point of fitting the equipment yourself in the first place. The object of this chapter is to show a basic tool layout, the use of which will enable the fitting of most pieces of ICE to your car. Most come to you courtesy of the author's toolbox, so if they look a little grubby and well-worn, I apologise! Wherever possible, look for tools to perform more than one function - the cordless drill is an infinitely useful tool, ideal also for those irritating odd jobs around the house, or even in the garden, where it saves having to trail 30 metres of extension cable around. When it comes to choosing which tools to buy, it is false economy to buy purely on the price ticket. A good quality item (and, therefore, usually one which costs a little more), is likely to last indefinitely and recoup its cost many times over. More importantly, by purchasing products from a reputable manufacturer, you will probably safeguard your health as well as that of your equipment. The stories of screwdrivers, pliers and spanners breaking under pressure are legion. Whilst being mildly humorous (as long as its someone else's tale) it could also be very serious. Skinned and bruised knuckles could be counted as a lucky escape. But the possibility of metal shearing and flying in the face of the operator (particularly the eyes) is not at all pleasant to contemplate. The moral, I repeat; buy good tools and they will look after you, in more ways than one.

Remember that your tools should always be kept clean and grease free in order to avoid contaminating any electrical contacts they touch.

*There's a story that many American ICE installers have a

Tools, techniques and workshop procedures

sign on the wall quoting labour rates. 'Systems installed for $25.00 an hour; $35.00 an hour if you watch, $45.00 an hour if you help!'. It's something to bear in mind.

The Black & Decker Workmate has been described as the DIYers 'third hand' and that's just about right. You can use it as a portable bench to keep your equipment off the floor and its clever design means that the twin 'jaws' will expand at all sorts of odd angles and so be able to hold almost anything secure. When used with the Kamasa portable vice, it's a home from home for the workshop dweller. These two are particularly useful if you have to work on your car away from your garage - they save a lot of trailing backwards and forwards. The Workmate folds flat when not in use and, if you want, can be hung on your garage wall so as to take less floor space.

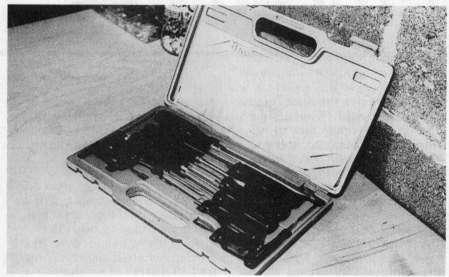

You can't manage without a screwdriver or two. This Teng set has all you'll need to fit your ICE and just about everything else. There's a range from large to small, crosshead and slotted. Whatever you do, use the right size and type of screwdriver everytime - not only can you easily 'chew up' the head of the screw in question, therefore making it impossible to remove, but also you run the risk of the screwdriver slipping and attacking your hands.

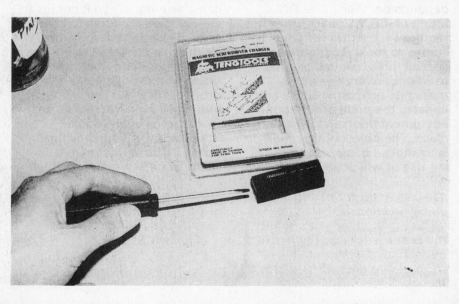

A magnetic screwdriver can be more than useful if you're trying to insert a screw in a tricky position. This Teng charger unit will charge otherwise 'normal' screwdrivers but remember to keep anything even vaguely magnetic away from your equipment and your cassette tapes.

Tools, techniques and workshop procedures

A lot of ICE fitting work involves removing pieces of interior trim - the sort of job ideal for the B & D power screwdriver. You don't have to do much trim removal and replacement before you appreciate this 'power to your elbow'.

Another essential item is a circuit tester - something which determines which are the power and earth leads. You could use a test light, like this 6v/24v Sykes-Pickavant model. Clip the crocodile clip on an earth and when the pointed end of the probe is placed on a live wire, the light will illuminate. Naturally, this works the other way around if you're searching for an earth. Alternatively...

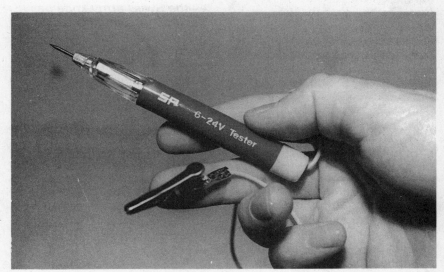

This tester performs much the same function, but with no wires to get in the way. When the probe is touched onto a live source, the tester earths itself through the operator and gives the message 'CIRCUIT OK'.

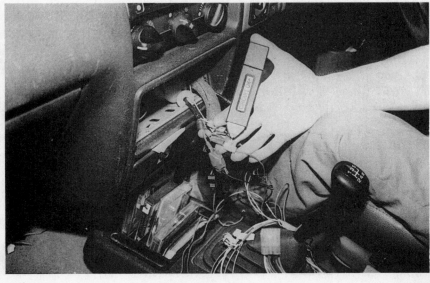

Tools, techniques and workshop procedures

If you're really taking the job seriously, invest in a fully fledged multimeter, such as this digital S-P variant. You're unlikely to find a situation that finds it lacking and given some good treatment, you'll still be together in 20 years time. Use it for finding exactly how many volts are present and for testing resistance and voltage drop, among other things.

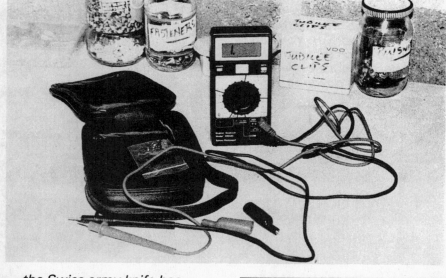

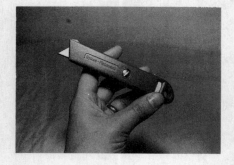

When you're fitting ICE, you'll often be cutting wires, trim, tape etc. A craft knife is a welcome addition, especially one like this, which has a retractable blade. Get into the habit of pulling the blade in after every single time you use it. You'll only kneel on it once in the open position to appreciate this wisdom here. Still on the cutting theme—

—the Swiss army knife has become legend, and quite right too. Whether you go for a multi-implement version, or something more simple, it will prove itself incredibly useful as you work on your car. This one has a permanent home in my glove box and has earned its keep many times over. However, all knives should be treated with extreme care. Bear in mind that a blunt knife is more dangerous than a sharp one - you'll find yourself applying more and more pressure to make the cut and eventually the knife folds under, or into, your hands - you get the message!

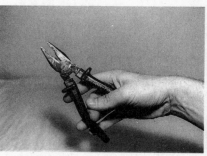

You'll need pliers - and not just one pair, either. Differentiate between this pair of Sykes-Pickavant combination pliers —

— and this pair of S-P radio (long-nose) pliers. Though still very strong, you'd use these for more delicate jobs and where their long jaws could reach where other pliers couldn't. Talking of accessibility —

16

Tools, techniques and workshop procedures

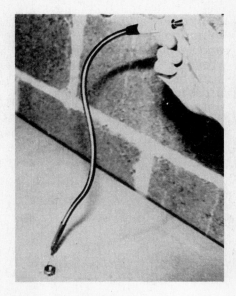

Once you're actually working on your car, you'll need somewhere to put the nuts/bolts etc. you've removed. Leaving them on the floor of the car is a sure way to be one or two missing when it comes to reassembly. The Teng magnetic tray is a smashing piece of kit - it's the bottom of the tray which houses a rubber mounted magnet and allows it to stick almost anywhere and hold those fasteners so you know where they are. Don't get it near your cassettes or cassette player, though.

—the S-P Pick-up claw is manna from heaven for the ICE fitter - if you can fit your system without dropping something fiddly into the complex innards of your car, then you're unique! Essentially, it's a strong, flexible rod, with a three-fingered claw on the end.

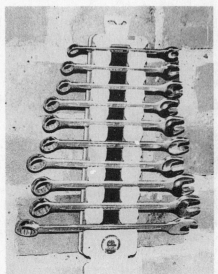

You won't need many spanners for basic ICE installation and so a neatly packaged set like this wall-mounted Teng collection is ideal.

You' need storage for bits and pieces; new connectors, screws etc., you've yet to fit and spare fiddly bits that may come in useful some day. You could buy a purpose made plastic container, divided up into different compartments or you could utilise containers which had a totally different purpose in another life. Manufacturers of marmalade, coffee and pate are just a few who have unwittingly contributed to my own collection. It's a good idea to label them clearly right from the start and don't forget that it's not wise to use glass containers as mobile carriers - these stay on or about my bench at all times.

17

Tools, techniques and workshop procedures

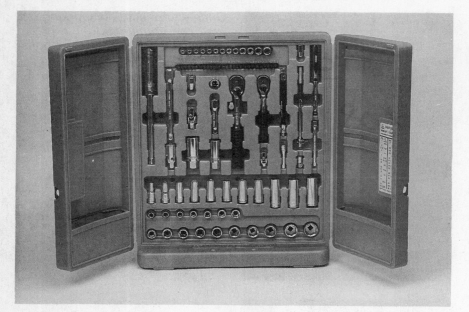

You could perform most motoring tasks with the sockets and ancillaries in this Teng Wallbox. There's a two sizes of socket set, with dozens of accessories, housed in this unbreakable case which can be mounted neatly on the wall.

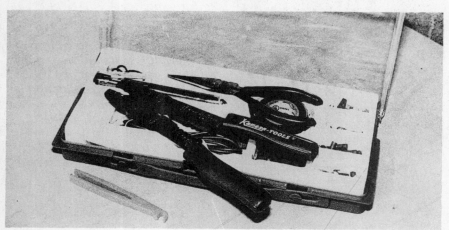

You'll need a crimper in order to make good electric connections. You can buy one separately or, as in this Kamasa set, as part of an electrical kit. This also includes a selection of terminals, long-nose pliers, and electrician's tape. Moreover, there's a strong plastic case to keep everything together.

You don't need to accumulate many tools before you start thinking about keeping them safely together. Having to open three different bench drawers to get the stuff you need is not likely to improve your patience! This S-P cantilever box is far from expensive but is of excellent quality and very strong. It has multi-layered trays, ideal for storing all those bits and pieces and when you're working on the driveway, it's much handier than trailing back and forth to the garage.

Tools, techniques and workshop procedures

Two kinds of tape are useful to the ICE fitter; electrician's tape can be used to make-up 'wiring looms' (where you have a multitude of wires heading in the same direction) and, of course, for insulating connections (note: insulating only, not securing!). Where you need to mark an otherwise awkward surface, a piece of masking tape makes it simple to use biro and when you need to drill a hole in your bodywork, it will prevent the drill from slipping.

For many ICE applications you won't need power tools at all, but when you do, it's likely to be a drill. A cordless model will prove worth its weight in gold when it comes to working in-car where the metal is usually fairly thin and you will often be drilling relatively delicate trim materials. This B & D 9032 drill has twin speed, hammer action (in case you get caught for a spot of household DIY), variable and reversible drive (making it ideal for occasional screwdriver applications) and both the chuck key and screwdriver bit are slotted neatly into a section in the top of the case. The rechargeable battery pack produces 7.2v which is more than enough for most in-car jobs.

Always use sharp drill bits, the right size for the job in hand. Most holes will require a pilot hole drilling first. An asset in any tool box is the S-P varicut drill bit, sometimes referred to as a Christmas Tree. Once the pilot hole has been drilled, you can enlarge to hole to any one of a number of preset sizes (the 'steps' around the edge of the bit) without the need to keep swapping and changing.

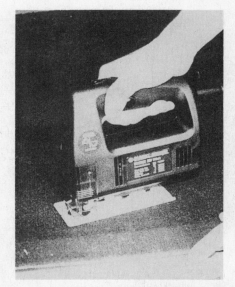

When you're fitting speakers, life is much easier when you can use a jigsaw. If you're completing a complex installation, you'll find it invaluable. Again, it can be used for many applications around the house and garden to justify its purchase. Please remember the safety precautions - used improperly, a jigsaw is a dangerous machine. This B & D saw features variable speed and a blade guard which should always be kept in the 'down' position when in use.

Tools, techniques and workshop procedures

Working under and/or in the dash of your car, you'll notice a distinct lack of light. Gunson's, 12v battery powered lead lamp is a good way to improve your vision. The metal cage protects the bulb from accidental harm and when replacement is required, it takes a standard car bulb and so it is not expensive. A variation on a theme —

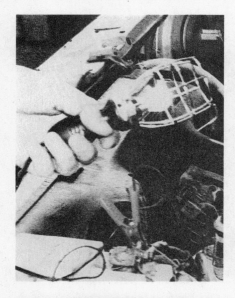

TECHNIQUES

—— is the Lighthouse. It has powerful, in-built nicad batteries which can be charged from the mains, or it can be operated from a vehicle cigar lighter. It's an ideal vehicle tool, as it has a strong single beam, a flashing red lamp (for emergencies) and a choice of single or twin fluorescent lights for when you're working in those tricky to get at places.

The Scotchlok must be the most misunderstood terminal in the world. To the professional installer, they are anathema; to the DIY layman, they are a godsend. When used correctly, it is a simple way to join wires; when used incorrectly, it is the source of much hair-tearing. Their use in-car is pretty much universal, and many manufacturers now supply equipment complete with a few in the fitting kit. You should, however, think twice about their use under the bonnet, where the hostile environment could cause problems. basically, the Scotchlok is —

So, you've spent all weekend up to your eyes in crimpers, wire and fuses, fitting the latest in hi-fi to your car. That moment you've been waiting for finally arrives; you switch on and...nothing. I know how you feel, and whilst destroying the whole car with your bare hands may seem to be the best solution, taking a 10-minute break with a cup of coffee and the instruction book is likely to be far more constructive. Today's equipment is notoriously reliable and efficient and a non-functioning unit is far more likely to be a 'silly' on the part of the installer (self included) than a set fault. Readers are privileged indeed to view the author's own personal mug, fully charged and ready for action. Over the years, it has been responsible for maintaining some semblance of sanity on many occasions!

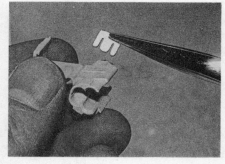

——a lump of plastic with a sharp metal cutter in its centre. Here the Scotchlok is seen opened out in a manner you wouldn't normally see.

Tools, techniques and workshop procedures

It is used mostly to piggy back one wire into an existing wire. The new wire is slid into one side of the Scotchlok and the original wire is slid into the other side.

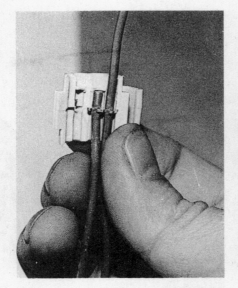

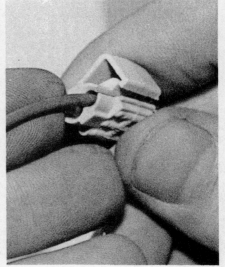

The cutter should be pressed down hard onto both wires. The object here is to get the cutter to pierce the plastic sheaths of both wires and thus make a connection. As it is not possible to actually see whether a connection has been made, it is very important to check each join as it is made. Incorrectly made joins can cause all manner of problems, especially if you've fitted the whole system and replaced all the trim.

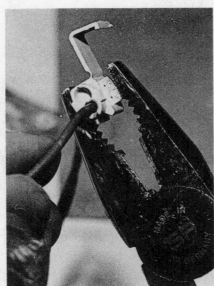

When the connection has been made, the outer lid of the Scotchlok folds over and clips onto the lower edge to prevent the top of the cutter accidentally shorting out on other wiring or earthing onto the car body. If this lid is damaged, throw the Scotchlok away and use a new one.

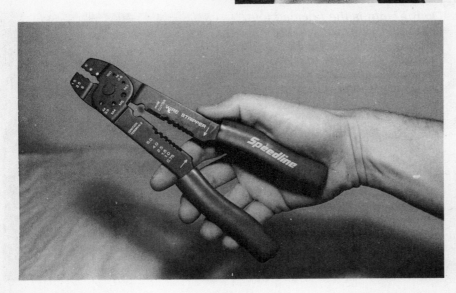

For crimping terminals, you'll need a crimper. They come in various shapes and sizes, but all serve the same purpose. Buy a good one - a cheaper one will soon show signs of strain and you'll find it not crimping the joins tightly enough.

Tools, techniques and workshop procedures

As with most crimpers, this one has the facility to strip the plastic sheathing from the wire. Note that the stripper is marked as to the different wire diameters. This allows you to strip only the plastic and not a handful of copper wire strands as well. When making a crimped join, you'll need to remove around 1/4" of plastic.

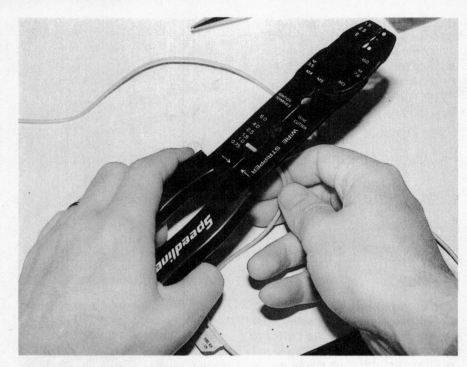

When crimping a join, remember that you should make two crimps - one to secure the wire securely to the terminal and the other crimp should be over the plastic sheathing to prevent the wire coming adrift should it receive a sharp tug. In this photo of a spade terminal, you can see that there are two edges which require crimping. (Note the insulator already in position.)

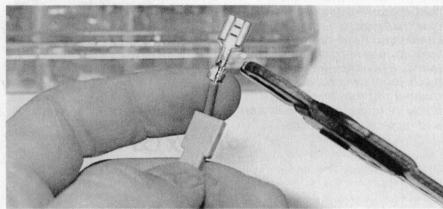

On the right is the finished join. On the left is a crimped join made on a terminal with a built-in insulator. Again, this shows the double crimping - on the sheathing and on the terminal and—

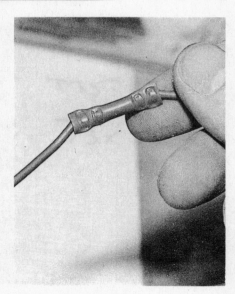

—the principle is seen again on this butt connector. These are used for in-line wiring joins. Note that there are two crimps at each join - not just one at either end.

Tools, techniques and workshop procedures

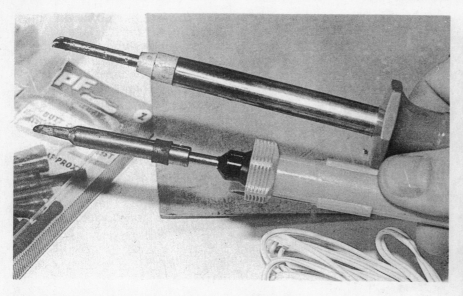

If you are soldering in-car, you have a choice of irons, all with pros and cons. Seen here are a 240v mains iron at top, and below a 12v iron. The mains unit will pack plenty of power to generate heat quickly. However, it will usually need an extension lead for ease of in-car working. A 12v iron, such as this S-P one, will run—

SOLDERING

The professional installer will almost always make his electrical connections by soldering. Note that an electrical connection should not be a *physical* connection and you should *never* let a soldered join take the strain. However, you must be (justifiably) confident in your abilities, for temperatures of 700 degrees or more are involved or you could end up with a claim under the 'fire' section of your insurance policy rather than some superb electrical joins. Regardless of the purist attitude, it is much better to make a really good crimped or Scotchlok join than a bad soldered one. If you are going to solder, take the time and trouble to practice - out of the car.

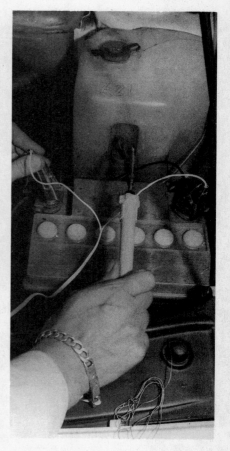

—from the car's battery and, though it still has a wire attached to it, is much simple to use and considerably more portable. For either of these electric irons—

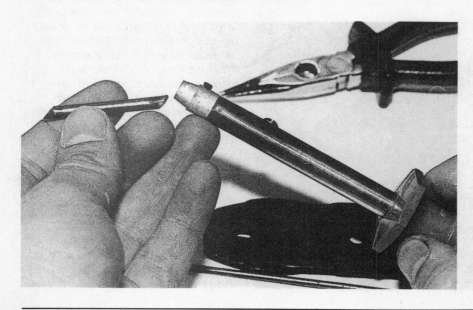

— there is a selection of different sized bits. The larger the solder join, the more heat you need and the larger the bit required. Naturally, you don't change bits unless the iron is cold! The third alternative—

Tools, techniques and workshop procedures

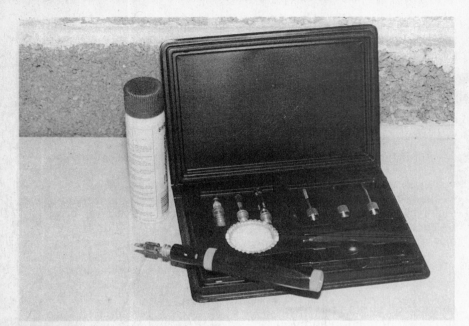

—is to use a gas powered iron. Seen here is the Kamasa kit, which also includes a number of alternative bits, meaning that the iron can be used for tasks other than soldering. Gas irons are powered by butane gas and are simple to refill in the same way as a cigarette lighter or portable hair curler. With no wires of any description, it is totally portable, though initially more expensive than the other two. Remember that the naked flame needs particular care inside a car.

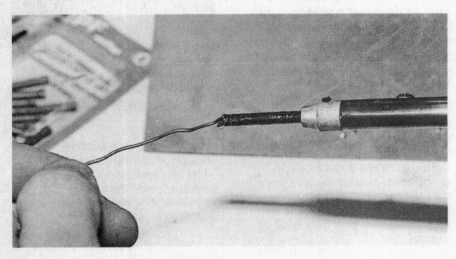

If you are using a new bit, it will need to be 'tinned' before use. This is simply a matter of melting some solder on the end of the bit, as shown here.

The art of soldering is easy in principle, though it needs practice to make perfect. Apply the iron to the join before the solder, heating it through. Then, when you touch the solder to the join, it should melt instantly and flow through both wire and terminal. Remove the heat and the solder will solidify almost immediately. Don't blow it, otherwise you could end up with a 'dry joint' - one which looks perfect on the outside, but which is loose in its centre and which can lead to all manner of intermittent faults. As you can see, it's far easier soldering (and making connections in general) when you're working outside the car and using the vice to hold the wire and terminal. A lot of people forget that the heat generated by the iron soon transfers down the wire to the fingers!

Tools, techniques and workshop procedures

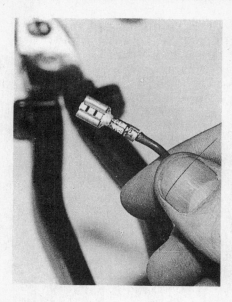

A good looking soldered terminal, if ever there was one. As you can see, the wire was crimped to the terminal first, in order to make a strong physical join - soldering is a way making electrical connections only. Don't forget that, unless you're inserting the finished item into a sealed plug, you'll need to fit an insulator to the wire before you make the connection.

SOME DOS AND DON'TS
Some dos—

DO take your time. A job rushed is seldom done correctly. Make sure that you have all the necessary parts, fuses and tools before you start work and that you have enough time to finish the job.

DO read the instructions! Lots of perfectly good equipment is ruined every weekend simply because people do not read the instructions properly - or at all.

DO use the correct tools for the job in hand, and make sure that they are the right size. For example, using a screwdriver of the wrong size or type to remove a speaker could result in a slip and a speaker with a hole in it.

DO disconnect the battery earth terminal before you start work on your car's electrical system. If you must have power to locate a live feed, do that first, then disconnect. Caution is required if your car relies on a steady 12v for such things as engine management memory. Suddenly disconnecting the power in such cases could cause problems, and you should check with the vehicle manufacturer beforehand.

DO make sure that the battery connections are tight and making good contact when you replace them. This is an ideal time to check for corrosion and/or clean them up.

DO connect ICE equipment into the fused circuit built into the car by its designers. Your handbook will tell you which fuse in the box you should use. By taking the feed from another source, you run the risk of introducing extraneous noise into the system. High powered systems use separate power feeds and fuse blocks, but these are usually expensive and inappropriate for the 'average' system.

DO be very careful if you are working on your car in the rain. Many owners do not have garages and so working in the rain (in the UK, at least) is a fact of life. Be extremely careful when using power tools - for preference use cordless and keep them dry.

DO use the right connections. There are plenty of different types commercially available and they're not expensive. Twisting two bits of stripped wire together and calling it an electrical join is not on - more than that, it's potentially very dangerous, even if it is covered with electrician's tape.

DO take the small amount of time and effort involved to route your wiring neatly. Tangled wiring not only makes it more difficult to trace a fault, but may also result in a deterioration of radio sensitivity and/or interference with both cassette and tuner.

DO make sure that all wiring is kept well away from sharp edges. Where wires are passed through a hole in the vehicle bodywork, make sure that a grommet is used. When routeing wire from the front of the car to the rear, make sure that it does not pass over any sharp edges and that it will not be chafed by any movement, for

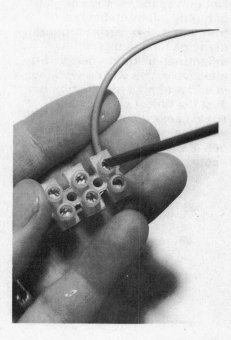

Terminal barrier blocks (or biscuit blocks) are another way of joining two wires together. Be wary here of overloading them, as they come in various sizes, depending on the amount of current they can take. Don't push too many wires into each hole - preferably one each side. By their nature, these blocks are bulky and can cause fitting problems if used at the rear of a radio/cassette deck. Always make sure that you use the correct size of screwdriver to wind down the screws otherwise you could chew up the heads and not be able to release them when required.

Tools, techniques and workshop procedures

example, the folding of the rear seats.

DO route the wiring away from the car's main loom otherwise extraneous noise could be introduced. Similarly, the DIN/accessory/battery and aerial cables should not cross for the same reason.

DO take great care when earthing any piece of equipment - it is possibly the most important connection you can make. Make sure that any earth lead makes a strong connection with the chassis of the car; it must be touching the metal and not just the paintwork. A bad earth will cause all manner of problems and no set will work at all if the earth is not connected. Where a fault is intermittent, it is worthwhile checking whether an earthing point has been subject to rusting. The flaking of rusted metal could be the cause of an occasional contact.

—-and some don'ts

DO NOT leave the keys in the ignition whilst you are working, even if you have the battery disconnected.

DO NOT force anything. A rule for DIY car maintenance in general, but applying particularly so with ICE, as the equipment you are dealing with is delicate. If you have a problem, check the instructions again. If you're still stuck, get in touch with the supplying dealer and/or manufacturer. Applying excess pressure could cause damage beyond repair.

DO NOT allow children in or around your vehicle, especially when it is unattended.

DO NOT touch any parts of the ignition system (especially the HT leads) whilst the engine is running. Few of us would do this deliberately, by if you're concentrating on solving an obscure wiring fault, you could find yourself doing it absent mindedly. The voltage generated is massive and far more than the mere 12v from the battery. The shock received from the HT side of the system could be serious to the point of being fatal.

DO NOT Leave unused plugs dangling free. For example, if your system involves RCA plugs and sockets, you may find yourself with some which you are not using at present. You should fit caps to them and tie them neatly out of harm's way.

DO NOT use small diameter cable where the job in hand requires something larger. Remember, the more current involved, the larger the cable should be. One possible result of using too small a diameter cable for, say, an amplifier installation, is a 'popping' noise through the speakers. Use speaker cable only for speakers, and for best results, use OFC (oxygen free).

DO NOT smoke when working on your car, and impose the same rule on anyone helping or watching.

FIRE

When you're working with vehicle electrics, there is always the risk of fire, however slight. Not only is the average car full of highly inflammable bits of trim, but it also carries around many gallons of highly inflammable liquid - petrol. So effectively, you will be working in and around a mobile bomb! Just the tiniest short circuit could cause a spark which, when combined with a whiff of petrol vapour could have very nasty consequences.

Apart from taking the obvious precautions, you should make it your business to equip your garage with a fire extinguisher (and, ideally, your car as well). In the garage, use a strong bracket and mount it on the wall where it can easily be reached in an emergency. In the car, the same principle applies, though space is always the problem. Don't mount it where it could come adrift in a dangerous manner in the event of an accident. Don't forget to read the instructions - it's no use trying to work out how to use it as your pride and joy burns happily before your eyes.

Chapter 2
Connections, Wire & Fuses

As in the business world, having the right connections makes you halfway to being successful. Without good connections, your equipment will either;
a) not work at all or
b) work badly or
c) be ruined altogether.

When connections are not made correctly, there could be several results. If the connection is poor, the operation of your equipment could be intermittent. If speaker leads touch while the set is connected, it could damage the amplifier and/or the speakers themselves. If the wires are 'live' they could well short out and give you a shock. Although seldom serious, it is not something you'd welcome. More importantly, a short circuit could result in a fire which, at best, may just damage your car, but at worst, it may damage you! One of the benefits of the rapid progress of technology is that nowadays, many of the connections which previously had to be soldered, can now be made using a simple connector.

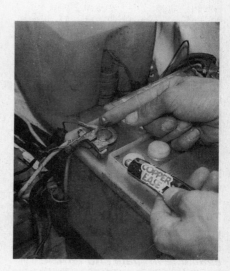

Before starting to fit any ICE items, it is a good move to check over the condition of the battery terminals, the earth in particular. They should be clean and bright, with no sign of corrosion. If they are not, they should be removed and thoroughly cleaned with a wire brush. To protect them against further corrosion, use Coppa Ease or Copper Ten, products which keep out moisture but retain the electrical conductivity.

Having cleaned the earth terminal on the battery, follow it down to where it is attached to the chassis or, in this case, the engine block. Deep down here, it is easy to overlook it and you'll do no harm by removing it and giving it the treatment already described.

Connections, wire and fuses

The biggest problem with connectors is that you never seem to have the right sort to hand when you need them. Buying a boxful like this makes a lot of sense – you get plenty of connectors to go at and with all those compartments, you've got a fair chance of actually finding the one you want. A box like this is well worthwhile, as you can use it indefinitely, stocking up regularly as you use up the various types of connector.

The most common connectors you are likely to meet are spade, or Lucar terminals. They are available in male and female versions.

When fitting any terminals, never leave them without some form of protective covering which will prevent the possibility of shorting out. The Lucar terminal at right has been fitted with a special plastic cover which slides over for protection. Don't forget to fit the plastic cover *before* the terminal - otherwise it won't fit! At left, is a crimped join (*double* crimped, note), where the crimped element is protected by the built-in insulator, but the terminal is left exposed. These would normally be used in cases where the terminal is to be fitted into an item such as a plug, where insulation is already present.

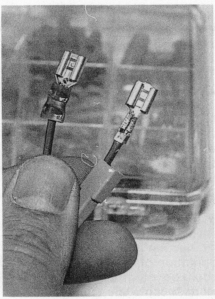

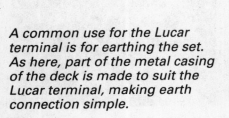

A common use for the Lucar terminal is for earthing the set. As here, part of the metal casing of the deck is made to suit the Lucar terminal, making earth connection simple.

Connections, wire and fuses

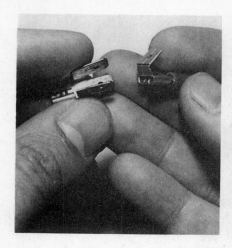

The 'piggy back' version of the Lucar terminals allows more than one wire to be connected safely to one connection. The one shown here on the right is a standard item, whilst the one of the left can handle a further three leads. If you ever use these, however, make sure that the fuse rating can stand two or more appliances being taken from the same source and that you are not overloading the wiring or the terminal.

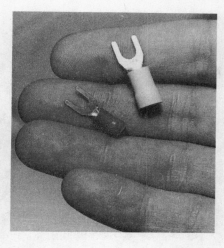

—the fork connector can be slid under the fastener in question. Again, the fork is usually used for earth connections.

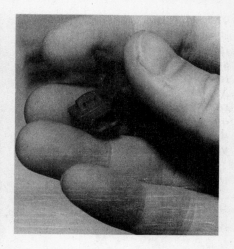

The Scotchlok connector is ideal for taking a feed from one wire to another. The two wires are placed into the clip and the metal grid is pressed down into position using a pair of pliers. The grid acts as a partial guillotine and cuts through the plastic sheath of both wires and thus makes an electrical connection. Their disadvantage is that the grid does not always make a good connection and so some care should taken when using them. See chapter 1 for more details.

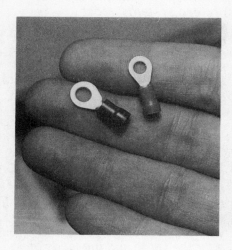

The eyelet connectors seen here are usually used for earthing purposes. They necessitate the removal of a nut of self-tapping screw, whereas—

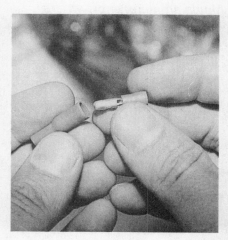

The bullet connector is well-named, as can be seen in this photo. As with the Lucar terminals, they too come in male and female and various sizes. The larger the connector, the more current it can handle, the latter being denoted by the colour of the built-in sheath.

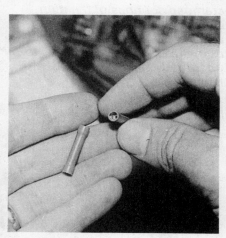

For an in-line join, you can use a crimped butt connector. The ends of the two wires to be joined are stripped of a little plastic sheathing and then placed into the butt connector. With either end of the connector double crimped, a good solid join is the result.

Connections, wire and fuses

The multi-pin DIN plug (this photo shows the 9 pin version) is usually used on uprated systems to link amplifiers etc., to other pieces of equipment. Most manufacturers can supply ready-made looms with plugs already connected on each end. It is possible for you to wire your own but—

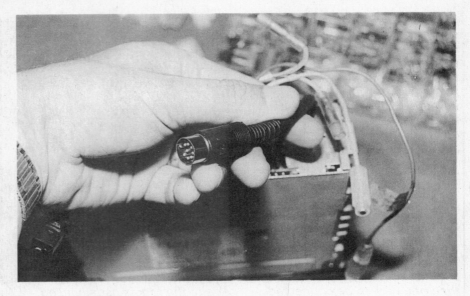

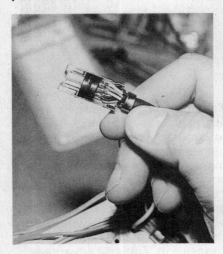

—as you can see, you will have to be more than proficient in the art of soldering in order to make good solid joins which don't short circuit on the others in the cramped confines of the plug. Unless you are really professional with a soldering iron, the advice every time is to go for the proprietary, made-up leads.

The sockets on the side of this Alpine amplifier show where such a plug would be used. Note, however, that the more usual RCA connections are also included, offering a choice to the ICE fitter.

RCA plugs are, again, used for uprated systems. Most amplifiers feature RCA sockets fitted which allow front end units to be connected easily to them. Again, you're well-advised to purchase pre-wired leads.

Many European radio/cassette decks have DIN speaker sockets fitted as standard. If your car is not fitted with speaker leads with suitable plugs you could alter the set to suit the leads or visa-versa. If you choose the latter, you can buy speaker DIN plugs, moulded onto the ends of short pieces of wire, making connection simple. Remember that the polarity of speakers must be kept constant, otherwise phasing will occur (see chapter 3)

Connections, wire and fuses

More pre-wiring. This plug and socket arrangement is from the back of a Far East radio/cassette deck. The large white multi-plug fits into the back of the deck and the four bullet connectors plug into the speaker cables. Note that for each speaker, there is a male and female bullet connector, the better to avoid 'phasing'.

FUSES

A fuse is the weakest link in any electrical chain. If there is a fault, the fuse will fail and disconnect the power from the equipment before it can be damaged. For this reason, it is important that the correct amperage of fuse is used.

The amperage fitted or recommended by the equipment supplier is chosen to protect your set at all times. A constantly 'blowing' fuse is a nuisance and there is a temptation to uprate it – don't! By fitting a fuse of higher amperage, you run the risk of the equipment being damaged before it blows. Compare the cost of a fuse (in pence) and that of your equipment, no contest. If you have this problem, then you need to sort out the cause, not the symptom. Conversely, by fitting a fuse of lower amperage, you may find that it blows every five minutes – not as expensive as the other option, but inconvenient.

Most of us have heard tales of fuses being replaced by just about anything metallic, from cigarette paper downwards. Except in very extreme cases (2a.m. in a howling gale in the middle of Dartmoor, for example) you should never, ever give this a second thought. Apart from financial damage (to your equipment) you run the risk of fire damage to your car and yourself. There are three basic types of fuse in common use, plus the fusible link type. The latter is fitted by some car manufacturers into a main supply cable, near to the battery. It is a highly rated item, designed to blow in the case of an accident and reduce the risk of fire. You should always carry at least one spare of each amperage in your fusebox and, when fitting any ICE equipment, you may need a whole pocketful – depending on your competence!

All cars have fuses. Before you start to fit anything electrical to your car, you must know where the fuse box is. In most modern cars, it is somewhere under the dashboard, though it can be found in the engine bay on the front bulkhead. Accessibility varies. In this photo, it is located under the dash on the driver's side. The square numbered boxes are plug-in relays.

Connections, wire and fuses

The three types of fuse in common use. At left is the ceramic fuse, used by older cars. The flat-bladed, cartridge, type is the 'modern' version of the ceramic and is easier to manipulate, but serves the same purpose. The glass fuse, at right, is usually found where—

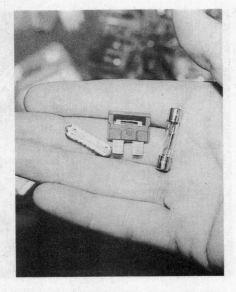

FLAT-BLADED CARTRIDGE TYPE

Amperage	Colour
3	Purple
4	Pink
5	Orange
7.5	Brown
10	Red
15	Blue
20	Yellow
25	White
30	Green

Ceramic type

5	Yellow
8	White
16	Red
25	Blue

Glass cartridge type – Conical ends

3	Blue
4.5	Yellow
8	Brown
10	Red/Green
35	White

Glass cartridge type – flat ends

2	Red/Blue
5	Red
8	Blue/Green
10	Black/blue
15	Brown
20	Blue/yellow
25	Pink
35	White
50	Yellow

—an in-line fuseholder is used, for example, in the power feed of a radio/cassette deck. The same rules apply to these fuses as to those found in the vehicle fusebox.

The fuses described in this table are all colour coded according to their amperage rating.

When a fuse blows, the visible indication is usually fairly obvious, as seen here!

Connections, wire and fuses

WIRE

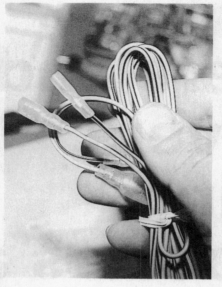

The table below shows wire sizes and current ratings.

Size	Current rating (amps)
9/0.012	5.75
14/0.010	6.00
36/0.0076	8.75
14/0.012	8.75
28/0.012	17.50

Metric	Current rating (amps)
9/0.30	5.50
14/0.25	6.00
14/0.30	8.50
21/0.30	12.75
28/0.30	17.00

Wire effectively forms the veins of your car's electrical system, but all wire is not the same, and it is important to use the correct type and size of wire for the task in hand. The wire used for power and earthing cable is usually made of up of a number of copper strands inside a plastic sheath. The rule with regard to power is simple; the more current involved, the larger the cable you need. Wire is graded according to how many strands of wire are in each sheath and on the diameter of each strand. Therefore, a cable marked 28/0.012 would have 28 strands, each with a diameter of 0.012". There are metric versions, where the diameter of the strands is given in millimetres. The usual size for fitting a radio/cassette deck (power and earth) would be 14/0.012 (metric equivalent, 14/0.03).

When you're wiring speakers, use speaker cable - ie, not just any old piece of wire you happen to find in your workbench drawer! Most speakers come complete with sufficient cable to enable their fitment to an average car. Note that this cable is already fitted with female terminals, one of which is larger than the other. These match the male terminals on the back of the speaker, a system which goes some way to eliminating phasing, a reason also for the 'striped' plastic sheathing on the cable.

Compare the 'normal' speaker cable on the right, with the upmarket oxygen-free (OFC) version. The latter is usually used with higher powered systems and well-qualified speakers, though, if you can afford it, it will do no system any harm. Because of the way the cable is made, it can carry a much better signal with less interference.

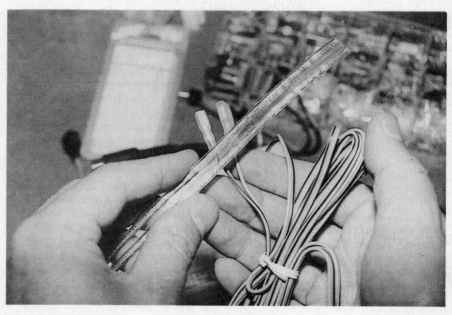

33

Chapter 3
Speakers and crossovers, including sub-woofers and sub-bass tubes

The importance of having suitable speakers in any audio system cannot be over-emphasised. All too often, they are regarded as an after-thought, even on complex and expensive set-ups. But what is the point of spending two month's salary on a state-of-the-art head unit and a pile of amplifiers then running them through the standard speakers? None, is the answer, and when considering your system you should start by looking at (and uprating where necessary) your speakers.

Basically, a speaker is a coil of wire attached to a flexibly mounted cone. The coil, which can move freely, is surrounded by a fixed magnet. The amplifier (whether a separate power amp or the one integral in the set) produces a variable alternating current which passes into the coil, generating a varying magnetic field. This interacts with the magnetic field produced by the fixed magnet and causes the speaker to move in and out, depending on the direction of the current. This in-and-out movement actually causes the air to move; the lower the frequency (or note) the more air there is to be moved. As such, speakers designed to reproduce very low, bass notes (woofers or sub-woofers) have to be much larger than those designed to handle only the highest frequencies - tweeters. The sound waves are produced to resemble the originals produced by the sound source (cassette deck, tuner etc.). Because the direction in which the cone moves is dependent on the current, the wires must be connected the same way round on all speakers in the system. If this is not so, then 'phasing' will result.

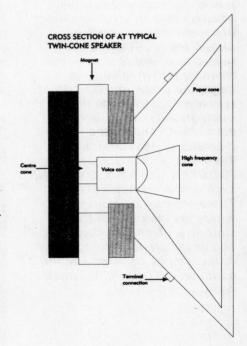
CROSS SECTION OF AT TYPICAL TWIN-CONE SPEAKER

Speakers and crossovers

This cutaway photo of a Clarion, triaxial (see later) speaker removes all doubt that a speaker is just a magnet with a cone on the front!

PHASING

Phasing is the strange sound effect produced when correct polarity is not maintained throughout the system. All speakers have positive (+) and negative (-) terminals, as shown on this simplified diagram. Speaker cables are coded in some way, usually by putting a stripe along one of the wires, in order to aid correct connection. If one speaker is connected in the right manner but another is not, then one cone will be moving out as the other is moving in. In this 'out of phase' state, the sound will become somewhat confused, as frequencies start to cancel themselves out. Bass response in particular will suffer. (Fig courtesy Harry Moss International)

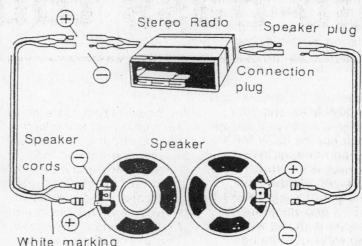

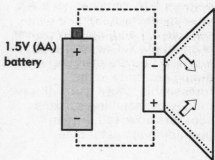

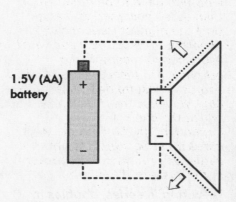

If you have a speaker with terminals which are not clearly marked, then there is a simple method of checking. Take an ordinary 1.5v (AA) battery and connect it across the speaker terminals. Take note of which speaker terminal is connected to the positive terminal of the battery. If the cone moves outwards, as here, then you have touched the positive terminal of the speaker to the positive terminal of the battery.

If, as in this illustration, it moves inward, then you have touched the negative terminal of the speaker to the positive terminal of the battery. Having ascertained which is which, you should mark them accordingly.

Speakers and crossovers

POWER RATING

It is vital to obtain the correct specification speaker to match the rest of your audio system. One of the most important aspects is power rating, which is measured in Watts (W).

A speaker, like an amplifier, will usually have two power ratings on its official specification. For example, on a speaker marked 10W/30W, the lower figure will be the RATED power and is the maximum continuous power that the speaker can handle. (This can also be shown as SINUS or RMS.) The higher figure is the MUSIC (or MAX) power and is the maximum power that the speaker can handle for short (sometimes very short) periods of time. These figures must correspond with those of the amplification available. The closer together the two power outputs are, the better the speaker.

For example, if a speaker can handle 100W MAX and 20W continuous, it's not as good as one which boasts 100W MAX and 60W continuous. When choosing speakers, you should look to exceed the quoted power output of the amplifier by around 50% and remember to compare like with like when it comes to ways of defining power. So, if your amplifier delivers 20W per channel MAX, your speakers should be capable of handling 30W power channel MAX. Consider also the future; if you are planning an amplifier uprate in the foreseeable future, you will save money by installing suitable speakers now, rather than upgrading at a later date.

IMPEDANCE

Impedance is a term used to denote the amount of electrical resistance. The unit of measurement is the OHM and speakers will usually be either 4ohm or 8 ohm. It is important to match your speaker impedance to that of your set. If you have a speaker impedance of, say, 8 ohms and the set is rated at 4 ohms, the speaker will lose some of its efficiency. However, if the impedance were mismatched the other way around, that is with the speaker impedance being the lower of the two figures, then not only will efficiency suffer, but there is also the possibility of damage to the unit. The wiring of the speakers can also result in differing impedances.

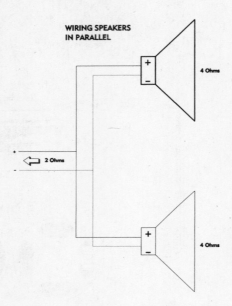

This diagram shows how wiring two speakers in parallel produces an impedance lower than the original, whereas—

—wiring in series, doubles it.

FREQUENCY RESPONSE

The frequency response of a speaker is, simply, a denotation of the highest and the lowest frequencies (or notes) that it is capable of reproducing. Frequency response is measured in Hertz (Hz) or thousands of kilohertz (kHz). Clearly, there has to be some overlap in terms of speaker capabilities; a tweeter will work between 2000 and 25,000 Hz, for example, but it will really only be at its best reproducing sounds in the 15,000 Hz-plus range. Speakers are divided into four main areas;

Tweeters – very high frequencies, typically rated between 2000 – 25,000 Hz.
Mids (or mid-woofers) – mid range frequencies typically rated between 300 – 12000 Hz
Woofers – low, bass frequencies, typically rated between 35 Hz - 4000 Hz
Sub-Woofers - extremely low frequencies, many of which are felt, rather than heard, typically rated below 200 Hz.

Because of the nature of sound waves, the lower the frequency, the less directional it becomes. As such, woofers should be placed towards the bottom of the car (in the lower door, of

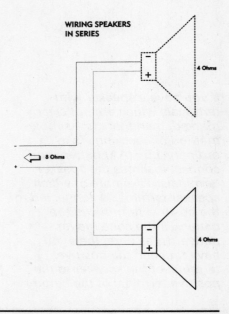

Speakers and crossovers

example) and tweeters towards the top – in the dash or even at the top of an 'A' pillar. The positioning of tweeters is particularly important as it is those which determine, to a great extent, the overall 'sound stage', i.e., the position of the sound in the car.

A sub-woofer speaker can, literally, be placed anywhere in the car and is usually found in the boot or hatch. When a vehicle comes with a sound system installed by the manufacturer, the speakers are usually mid-range versions, capable of giving a reasonable sound across the spectrum. However, they will tend to be lacking at the extremes - bass and treble. Some cars come with a mid/tweeter arrangement, which gives a much crisper, cleaner sound, though bass will still tend to be lacking somewhat.

It is often the case that owners spend lots of money on replacing the head unit and/or amplifiers without giving much thought to their speakers. This is the audio equivalent of turbocharging your 1.3 litre town shopper but leaving it still running on standard issue tyres and brakes - not wise!

Many vehicle manufacturers position their standard fit, wide-range speakers atop the dash, as seen here. As such speakers are usually only about 3´" in diameter, there clearly isn't going to be much in the way of bass response, a problem exacerbated by their positioning - bass speakers need to go lower down in the car. Whilst you can uprate these speakers quite simply (most ICE manufacturers produce a wide range of 'custom' speakers, designed to be a straight swap) you'd be far better off buying a component speaker system, fitting the mid/woofer in the door and using the original speaker position for a tweeter, a task for which it is infinitely better suited.

This is the sort of position you need to try and find for your mid- woofer. In this particular car, the door trim has been reworked with an MDF (medium density fibreboard) baffle to hold the speaker securely in place.

Speakers and crossovers

A typical standard-fit speaker, in this case a 6" x 4" elliptical model. It will reproduce the middle frequencies reasonably well as long as you don't ask too much of it. However, it's not hard to find fault with and you'll be missing lots of treble and bass frequencies.

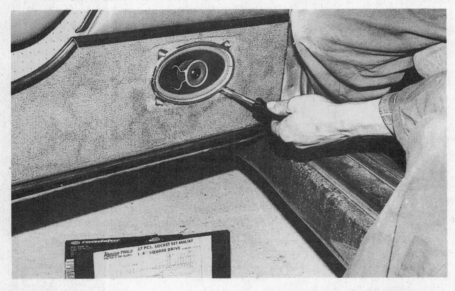

The simple upgrade path is to replace it with a speaker of the same size but of the coaxial variety - two speakers within the same housing. The smaller speaker in the centre is a tweeter, designed to reproduce the higher, treble frequencies. Though it is an easy swap to make, having the tweeter so low down in the car is not really the best place.

The coaxial speaker should not be confused with the twin cone speaker. As seen in this photo, there is an extra, small cone at the centre of the speaker and, whilst it does tend to give a cleaner sound, it is not as good as having two speakers.

Speakers and crossovers

One can divide up the workload still further by constructing an enclosure containing three speakers which share out the responsibility of reproducing the frequency range supplied by the head unit. This Philips triaxial is a hefty 6" x 9", elliptical version, though—

—this Roadstar, 5"" is another variation on the theme.

Another way to run three speakers on a single channel, is to use an enclosure such as this one. The obvious drawback here is the space required and the fact that it would be very much on show (inviting a possible break-in). As they would usually be mounted on a rear shelf, hatch-back owners should remember the rules with regard to routing wires. It's very easy to trap speaker leads and, should the wires touch, it is possible to damage the output stage of your amplifier.

COMPONENT SPEAKER SYSTEMS

As already mentioned, the more speakers you split the sound between, the better the quality of sound you get. Many companies now produce component speaker systems - two or more matched speakers supplied with a suitable crossover network. A mid-woofer and tweeter is the most common option, though there are many variations on this theme. Mounting a mid-woofer in the lower door and the tweeter atop the dash/in the upper door, is something you will find in many vehicles. It's not that hard to achieve and you will get a much better level of sound than can be achieved by just a single speaker in the door panel or, worse, in the top corners of the dashboard.

Speakers and crossovers

Mounting positions of component speakers are usually dictated by the size and shape of the door, though this installation is typical and basically correct, with the mid-woofer as low down as possible with the tweeter much higher.

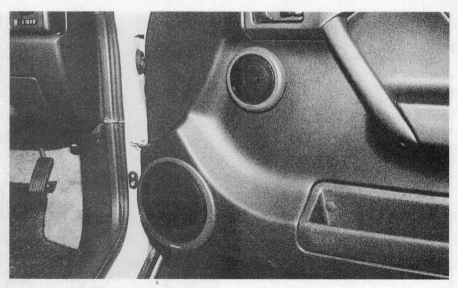

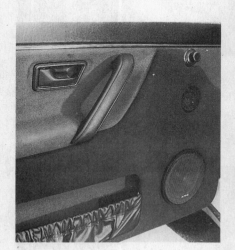

This car utilised a special replacement panel made specifically by Connaught Designs. It is made of strong plywood and facilitates the fitting of a woofer and a tweeter. The panel makes an excellent baffle and ensures that the sound comes straight out of the speakers into the car and not back into the inner door! In this case, the speakers are Alpine complete with—
(Courtesy Connaught Designs)

If you don't want to cut your dash or if there isn't a suitable mounting place, you can still put your tweeters in the right place (as high up as possible) by sticking them onto the dash, as shown here. Hiding the speaker cable is always the tricky part and often, the best you can hope for is just to keep as much of it as possible out of sight.

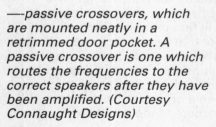

—-passive crossovers, which are mounted neatly in a retrimmed door pocket. A passive crossover is one which routes the frequencies to the correct speakers after they have been amplified. (Courtesy Connaught Designs)

Speakers and crossovers

Give some thought as to the outer appearance, and not just for aesthetics, either; a set of highly expensive speakers on display make a high security risk and point to the possibility of there being yet more expensive equipment elsewhere in the car. This very unobtrusive, but expertly trimmed speaker panel hides—- (Courtesy Dards Electronics)

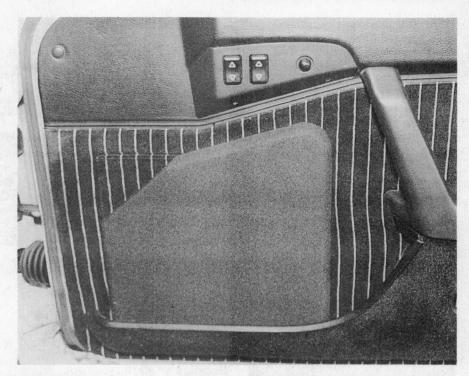

—-a mid-woofer/tweeter combination, housed on a purpose-made baffle. The baffle has been designed so as to require the minimum of door panel cutting (just a few millimetres on one side) and sprayed mat black to suit. It would be tempting to leave it on show, but not awfully wise. In order to ensure that best performance was obtained—- (Courtesy Dards Electronics)

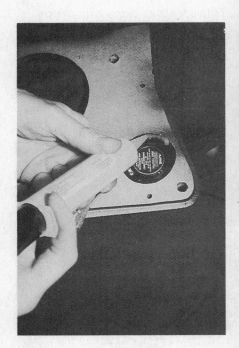

—-the push-fit tweeter was sealed into position using silicone sealant. (Courtesy Dards Electronics)

CROSSOVER NETWORKS

Once you start improving your speakers, you'll run into the crossover network. In effect, a crossover is an audio 'points policeman' designed to sort out which frequencies should go to which speaker. Some units are described as active and some as passive.

Active crossovers use active components such as transistors and integrated circuits. They are used to sort out the frequencies before they are amplified. As such, it means that you will need an amplifier for each frequency separation.

Passive crossovers use passive components, such as resistors and capacitors and are used to separate frequencies after the amplification stage.

Speakers and crossovers

A good crossover will offer a way to vary the points at which the frequencies are switched. As you can see here on this Altec Lansing ALC11, there are a variety of settings to be selected (using a small screwdriver) to produce the sounds that suit your ears best. As crossovers are (or should be) mounted well out of harm's way, it's something you do when the speakers are first fitted and then leave well alone. (Courtesy Altec Lansing)

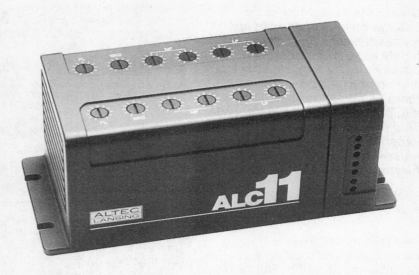

Mounting crossover units often requires some ingenuity, as they have to be out of harm's way be reasonably accessible because of all those plugs. This one is mounted in the boot of a saloon, alongside the bootlid mechanism.

This complex Sony unit drops nicely into the rear nearside cubby hole, mounted on a specially made board.

Speakers and crossovers

A PASSIVE CROSSOVER - DIVIDING THE FREQUENCIES AFTER AMPLIFICATION
A TYPICAL EXAMPLE, WHERE A 2-CHANNEL AMPLIFIER WOULD FEED FRONT COMPNENT SPEAKERS, SPLITTING THE SIGNAL BETWEEN A TWEETER AND A MID-WOOFER

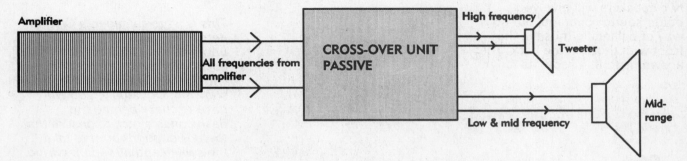

This diagram shows a simplified application for a passive crossover unit, in the way you would find in many 2-speaker component speaker systems. It shows the amplified frequencies entering the crossover and then being split up into high frequencies (which go to the tweeter) and the rest (which go to the mid-woofer).

AN ACTIVE CROSSOVER - DIVIDING THE FREQUENCIES BEFORE AMPLIFICATION
A TYPICAL EXAMPLE WHERE A 4-CHANNEL AMPLIFIER IS USED TO POWER TWO REAR COAXIAL SPEAKERS WITH TWO CHANNELS BEING BRIDGED TO POWER A SUB-WOOFER SPEAKER OR SUB-BASS TUBE

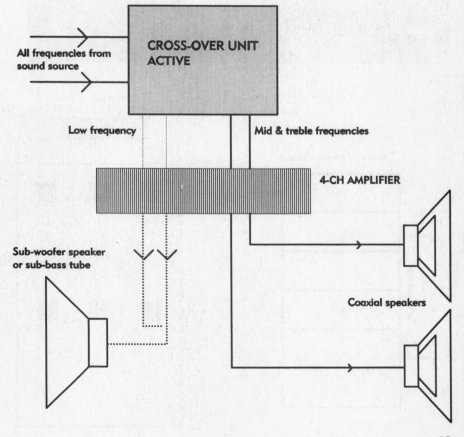

This is a typical application of an active crossover, where all frequencies enter the unit before the signals are amplified. The crossover divides up the various sounds and then passes them into the amplifier, to be passed on in their boosted state to the respective speakers. It is common to find this sort of usage at the rear of a car, where two of the channels (of a 4-channel amplifier) are bridged to power a sub-woofer and the other two are used to power a couple of coaxial speakers as 'rear fill' for a front component speaker system.

Speakers and crossovers

FITTING A FADER

Most modern sets come ready-wired for four speaker operation and have an integral front to rear fader. However, if you have a two channel set, then adding two speakers will give you better sound coverage, but no way of balancing those in the front with those in the back. The answer is to fit a separate fader.

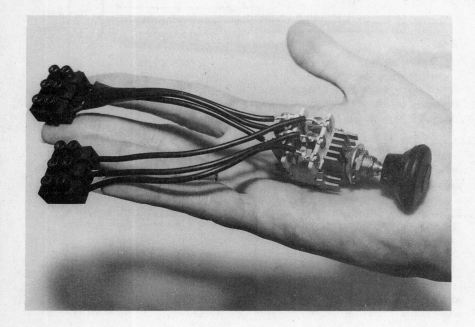

This is a typical fader, a Clarion item, designed to fit unobtrusively into the dashboard or centre console with just the adjustment knob showing. Obviously, you will have to check behind the dash/console etc., to ensure that there is enough room to take the switch and attendant wiring. The wires from the four speakers could be connected in various ways, with soldering being the favourite, as long as you are fully proficient in that particular skill. Alternatively, you could fit bullet connectors to the wires to match those already fitted to many speaker leads. Finally, you could do as shown here and fit the leads from the fader with terminal barrier blocks, a method which requires nothing more than a small screwdriver (not too small, though, or you'll chew up those delicate screw heads).

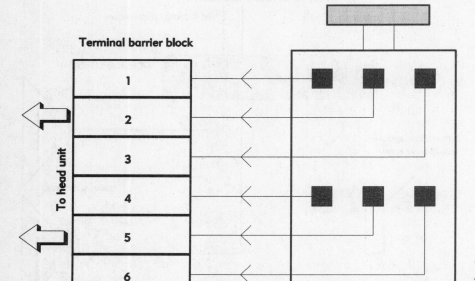

This diagram shows how the fader terminals should be wired to the terminal barrier blocks and...

Speakers and crossovers

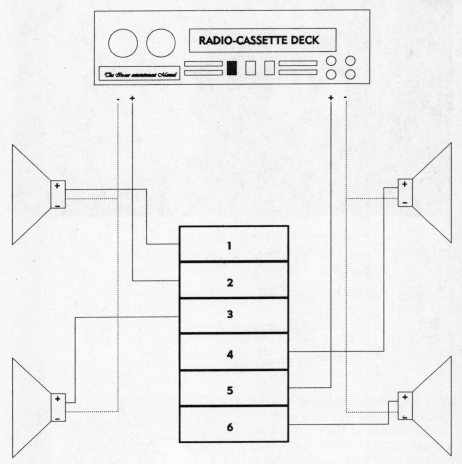

... this one shows the connections required for the speakers and the set. You must take great care to get these right, as not to do so could cause phasing, or if the wires are shorted together, damage to the output stage of your amplifier.

SUMMARY
When purchasing speakers, you should bear in mind three main points with regard to the technical specification.
a) The power handling capacity should be at least equal to that of your set, preferably with a 50% safety margin.
b) The impedance should match that of your set and
c) The frequency response should be at least as wide as that of your set.

If you have any doubts as to the suitably of your speakers with regard to the performance of your deck, the ask the dealer in question. If he can't give you an answer, you're probably at the wrong dealer!

SPEAKER FITTING
The 'before your start' checklist
1. First and foremost, buy the best you can afford. Even wait a while until you can afford those you thought you couldn't – if you see what I mean!
2. Plan ahead. Uprating your speakers on a regular basis to suit similar upgrades in other equipment is an expensive business.
3. Make sure that the speakers you buy will match your system.
4. If you uprate your system, uprate your speakers as well. There's no point in doing one without the other. You're excused if you planned ahead and have suitable speakers already.
5. If you have a CD player, you must have speakers to suit. The extra dynamic range and pure ability of CD means that 'ordinary' speakers will be left sadly lacking.
6. Make sure that your new speakers will physically fit in your car before you buy them.
7. Ensure that your speakers are held securely in place. If you are mounting in existing trim, make sure that at least one of the speaker mounting screws passes into metal, in order to prevent the speaker from vibrating.

The standard fitting place for front speakers is in the doors; for the most part, there isn't much choice! Before you buy your speakers and start hacking away with a craft knife, you should remove your door trim panels and see what lies beneath. Items such as the windows and armrests etc., could well foul on the back of a speaker and prevent correct installation. Some cars have a 'hole' in the pressing designed to accommodate a speaker magnet.

When it comes to positioning a speaker, it's often a compromise, for, as you will remember, you really need speakers which deal with high notes, high up in the car and visa versa for the low notes. If you only have one speaker (either a wide-range single unit, or a coaxial) then something has to give. In general, you will probably be governed by the door configuration.

Speakers and crossovers

A nice flat door panel is manna from heaven for the speaker fitter. In this case, a hole has been cut to take the speaker and the speaker has been self-tapped into place. Note that at least one of the self-tappers should find its way into metal, as seen—

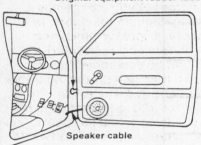

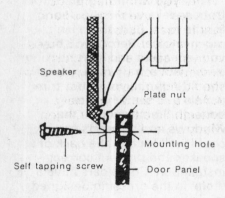

—in this diagram. If you secure the speaker to the thin card trim, you are risking excessive vibration and loss of bass frequencies. Try to get at least one of the securing screws (and preferably all of them) into some metal. Wherever you screw into card trim, don't forget to use the metal plate nut so that the self-tapping screws have some metal to hold onto. (Courtesy Harry Moss International)

With door mounted speakers, the cables have to be routed from the door to the car. This is a good opportunity to ruin your ICE system! In some cars, there will already be a suitable hole in the 'A' pillar and door - either to take existing wiring, such as central locking and electric windows, or in anticipation of same. If this is the case, there will probably be a rubber tube protecting the wiring as it passes from door to car. If not, and there is not such a thing with your speakers, you should buy one before you go any further. If you have to make your own holes, drill the one in the door slightly below the one in the 'A' pillar and always fit a grommet, so that the wires do not chafe and short circuit. Don't forget to leave some slack in the wires to allow for the door to open to its full extent. (Courtesy Harry Moss International)

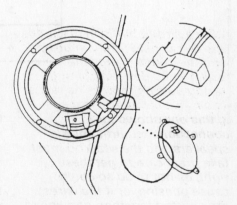

A good speaker will have a cable clip built-into the back. With the terminals in place, a small loop of cable should be allowed and then the clip pressed down hard onto the cable. Should the lead then receive a sharp tug, the terminals will remain in place. (Courtesy Harry Moss International)

Speakers and crossovers

SPEAKER MOUNTING WITHOUT BAFFLE

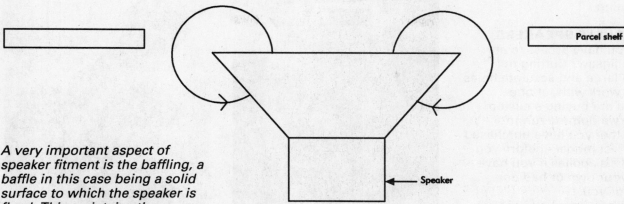

A very important aspect of speaker fitment is the baffling, a baffle in this case being a solid surface to which the speaker is fixed. This maintains the pressure either side of the speaker and is usually used where the main fixing area is flexible, for example, the rear shelf of a hatchback. In this diagram, the movement of the speaker and the shelf is sufficient to cause the sound to move back around on it self. This would lack bass response and give a distorted, muffled sound.

With a strong baffle in place, the speaker is held firmly in position and the sound leaves the cone as it should do, straight out, taking no side roads. The result is a much cleaner sound and greatly improved bass response.

FITTING FRONT SPEAKERS

In general, fitting speakers in the front of your car means fitting them in the doors. In theory, a speaker producing bass frequencies (woofer) should be positioned lower down than one producing treble frequencies (tweeter). In practice, it is often not possible; if you have a coaxial speaker (two speakers, usually a tweeter and mid-range, in one enclosure) then, by definition, some compromise has to be made. Moreover, the design of the inner door is likely to force your hand as to where a speaker is mounted. Clearly, unless you are particularly skilled in the art of cabinet making and can reconstruct your door trim panel to suit, you're stuck with some concessions.

The 'before you start' checklist
1. Make sure that your speakers are complete. Have you the necessary screws and terminals to effect the fitting.
2. Ensure that your speakers are technically suitable and physically capable of fitting where you want them to go.
3. Fuses - have three on hand just in case. Speakers themselves will not need fuses but you could easily disturb something on your head unit in the fitting process.
4. Make sure that you have enough time to finish the job. Many speaker fitments require

SPEAKER MOUNTING WITH BAFFLE

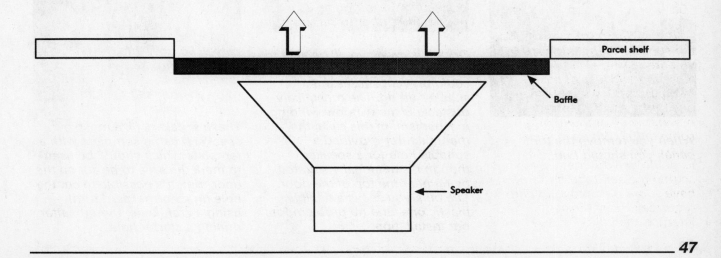

Speakers and crossovers

plenty of trim removal and replacement - always time consuming.

REAR SHELF SPEAKERS
5. Do you have access to an electric jigsaw? Cutting out several large and accurate holes is hard work without one.
6. If you are buying a custom shelf, try a dummy run to ensure that you have purchased the correct model - before you cut it! This applies if you have made your own or had one made for you.
7. Make sure you have enough speaker cloth for the shelf and that you have either a staple gun or glue gun to secure it to the shelf.
8. If you're going to take the shelf out on a regular basis, it's a good idea to fit a couple of plugs/ sockets for the speaker leads, to make life easier.
9. Tools for the job. A typical speaker fitment will require;
Crosshead/slotted screwdrivers
Crimper/terminals
A few suitable terminals
Tape measure/pencil
Masking tape
Craft knife
Where it is necessary to cut trim and/or rear shelf fitment:
Jigsaw
Electric drill/bits
Stapler/glue gun

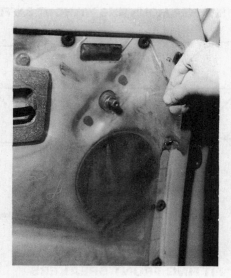

Important, as it keeps any moisture away from the trim panel and the inside of the car generally. You'll need to ease it away very carefully, pulling gently at the edges.

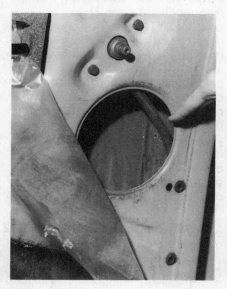

Once it is away, you'll be able to see any obstructions which could prevent successful speaker fitting. The most likely obstacle is the window winding mechanism. In this case, the manufacturer provided a suitable hole for a speaker, though it's not ideally situated, being near the top of the door. The pipe you see is a flexible plastic one and no problem for our installation.

When you know where your speaker is to fit, it's more than a good idea to offer up the speaker grille to make sure that it will cause no obvious obstruction. Will that window winder clear? Even if it does, it's possible that it will trap your fingers everytime you use it. Check things like this before you start drilling and cutting.

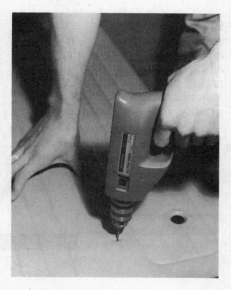

These speakers (like most speakers) came complete with a template which should be used to mark the area to be cut on the door trim. It's possible to cut the hole (in a normal card trim) using a craft knife, though, after drilling a starter hole...

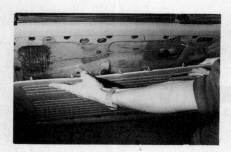

When you remove the trim panel, you should find...

Speakers and crossovers

... it's much simpler with a jigsaw

Having mentioned moisture inside the door panel, you'll appreciate the need to fix the plastic water deflector to the speaker before fitment; make sure you get it the right way up!

The speaker cable has to get from the inside of the car to the door. Most cars have holes pre-drilled in the door jamb and the door (some will already have electrical cables running through them, for accessories such as electric windows and central locking). If you haven't got any holes, you'll have to drill a couple. Make sure that the hole in the jamb is slightly higher than the one in the door, that you use plenty of rust proofing agent and that grommets are always used in both holes.

You should always aim to get at least one of the speaker mounting holes through metal as well as the card trim. The more secure the speaker, the better the sound you will get. In this case, the size of the hole was perfect and so four holes could be made in the metal. All, of course, were rust proofed. Note that the plastic covering inside has been cut to suit the speaker, but has been left in place around the edges to prevent water damage.

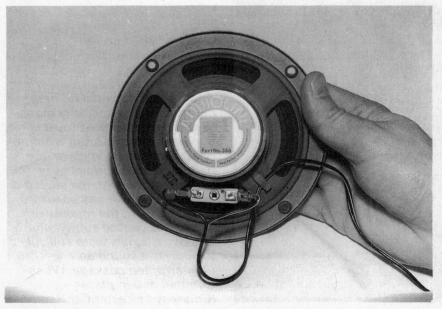

Having routed the wiring to the inside of the door, don't forget to push the terminals onto the back of the speaker (don't laugh; it's a common mistake!). Make sure that the polarity is right, to avoid phasing, and if you have a cable grip, as seen here, use it. That way, it will prevent the terminals being pulled off and/or possible speaker damage, should the cable receive a sharp tug.

Speakers and crossovers

FITTING REAR SHELF SPEAKERS

Fitting speakers in the rear shelf is common practice. If you have a saloon, you start with a much more stable base for speaker mounting. However, you should be very wary of cutting large chunks of metal from the structure of your car in order to mount speakers. By doing so you could be taking away much of the car's structural rigidity which could be vital in the event of an accident.

In hatchbacks, the vehicle manufacturers generally favour mounting speakers in the shelf supports at either end of the shelf. There are various other positions, but, to date, no one actually uses the shelf itself. Clearly, that flat shelf area cries out for speaker mounting, but it's not as simple as that. There is the routeing of the wiring to be considered and the effect that the weight of the speakers will have. Overall it's good advice to remove the original shelf (and keep it safe for when you sell the car) and replace it with an uprated version. You could either fabricate one yourself or fit a custom-made model, such as the Auto Acoustics version shown in this fitting section. Remember, also, that if you are fitting some particularly large speakers, you may need to strengthen the shelf supports to take the increased weight.

Rather extreme, but effective. The rear of this Sierra has been totally given over to speaker location. Clearly, the rear shelf is a specially made item and holds four sub-woofers. Alongside are 3-speaker component systems. This is a professional demonstrator and represents a hefty investment in terms of cash and time - not least as the suspension had to be beefed up in order to take the extra weight! (Courtesy Braybrook Car Radio)

Another professional fit, here. In this case, the specially made section was designed to sit in the rear hatch area, but with the original shelf still in place, the better to divert curious eyes elsewhere. Though the shelf did not actually have any speakers mounted on it, the respective holes (one central sub-woofer and two triaxials were still cut, to allow the sound an uninterrupted passage. When trimmed and in place— (Courtesy Leicester Car Radio)

Speakers and crossovers

—you wouldn't know that this was a standard car. Not particularly good for the ego, but great from the security point of view and there's nothing that deflates the ego like have some maniac rip all your stereo gear out with a crowbar and a hacksaw! (Courtesy Leicester Car Radio)

When it comes to DIY fitment, this is the kind of thing you're likely to find. It's undoubtedly an easy way to fit rear speakers and, given that the speakers are not so heavy as to 'bow' the shelf and that they are fitted correctly, they're OK here. However—

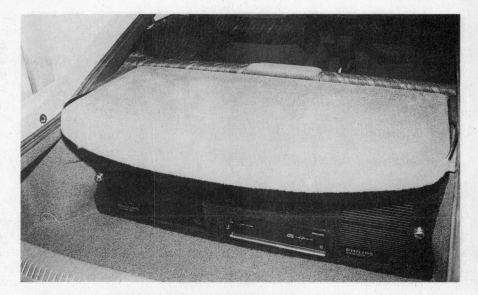

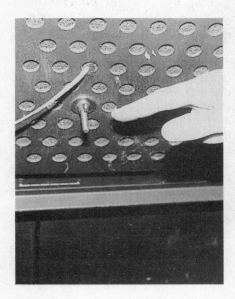

—there were problems with these particular speakers. The securing stud is deliberately made longer than necessary by the manufacturers so that it will suit a wide range of applications. The thinking is that you will shorten it to suit - not leave it like this, where it could easily catch on luggage or your person as you load it. It needs shortening and some kind of protective cap on the end - try 'shrink wrapping' (gently heating) a piece of old cable sheathing.

The problems didn't stop there, though. The routeing of the speaker cable was appalling, with no attempt being made to take it safely from rear to front of the car. As seen here, the cable is just hanging down, waiting to snag on anything and everything—

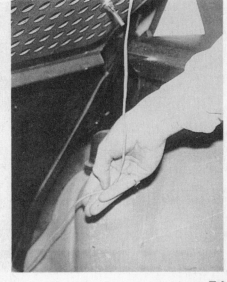

Speakers and crossovers

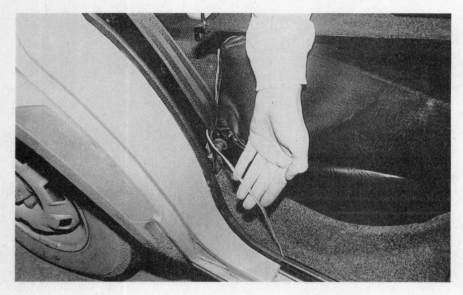

—-and look at this; clearly, it could easily be damaged by rear seat passengers and it's a wonder it hadn't been ripped free already by a combination of seat belt and folding rear seat.

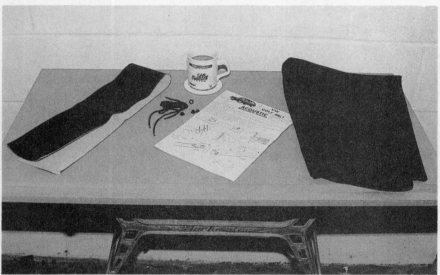

Well, that's how it shouldn't be done, now let's see how it should be done. Dards Electronics of Milton Keynes went through the, relatively, simple process of uprating rear shelf speakers. They took it in two stages, fitting a new shelf and then the speakers themselves. This shelf is one of the Auto Acoustics range of custom made units designed to suit individual models exactly. It is made of MDF (medium density fibreboard) which is very much industry standard as a material for securing speakers. It comes complete with a generous portion of acoustic speaker cloth, leatherette trimming where necessary and eyelets and shelf supports. The coffee, as you may have guessed, is a personal optional extra!

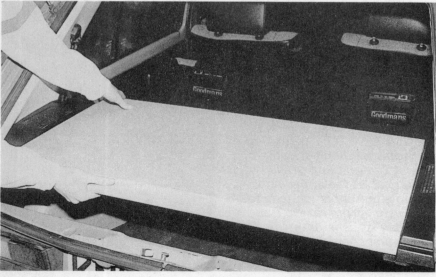

Auto Acoustics recommend that the first task is to hold the new shelf in place to ensure that it fits as it should. If you've ordered slightly the wrong shelf (different year or model etc.) then you need to know before you start making holes in it!

Speakers and crossovers

The speakers being fitted to the shelf were a set of Philips PRS602 component speakers, a 6" mid-woofer and a tweeter. The size of them makes them highly inappropriate for a standard shelf, which would simply bend itself in two trying to bear the weight. As you can see, everything is included in the Philips box, including some rather nice OFC (oxygen free) speaker cable and the two passive crossovers required to split the frequencies and send them to their respective speakers. Also, the speakers come with protective grilles, required if you are 'surface' mounting, but not needed in this case.

Before you can fit the speakers, you have to mark out their exact positions, which is where most DIYers go wrong. A lax attitude or rushing, due to a lack of time can result in real mess; a shame, because getting it right takes little more time and effort. Use the templates supplied and offer the speakers up to ensure that everything looks right. Don't forget that the lower the frequencies (and the larger the speaker) the less directional they become. Therefore, place the tweeters and smaller speakers near the edge and the woofers etc., nearer the centre. Placing the larger speakers toward the front of the shelf will prevent them being a nuisance when you open the hatch to remove luggage. It pays also to consider your plans for the future. In this case, we allowed for a possible future fitment of one or two sub-woofers in the centre of the shelf.

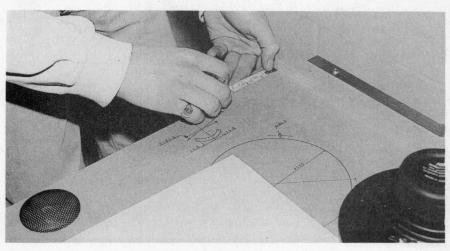

Yet more measuring - you cannot be too careful at this stage; measure twice, cut once, should always be your motto!

Speakers and crossovers

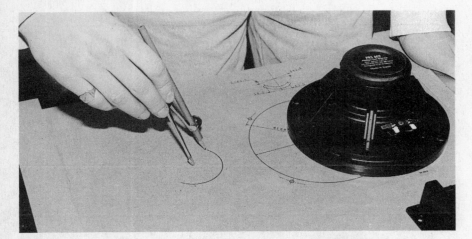

Now you'll wish you'd hung onto those old school compasses. The positions of the mid-woofer, tweeter and crossover have to be marked, bearing in mind where the wiring is to be run.

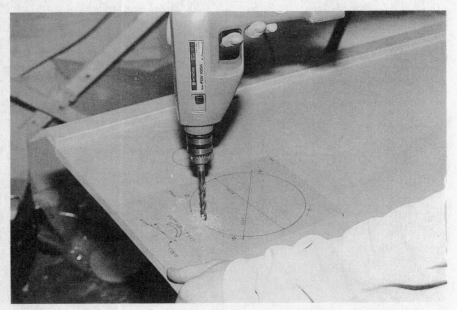

Too late for second thoughts. In order to use a jigsaw (highly recommended unless you feel you need the exercise) a pilot hole has to be drilled to form—

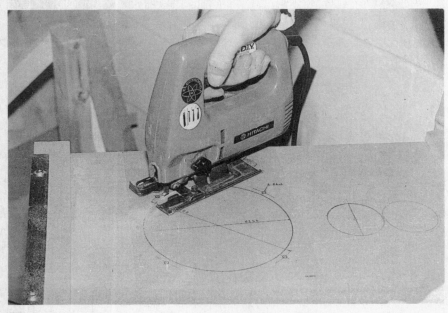

—a starting point for the jigsaw blade. Take your time and concentrate on keeping that blade exactly on the line. Eye protection is recommended.

Speakers and crossovers

Perfect symmetry and just half an hour to do it. Use a half-round file to smooth off the edges to prevent them snagging on the speaker cloth.

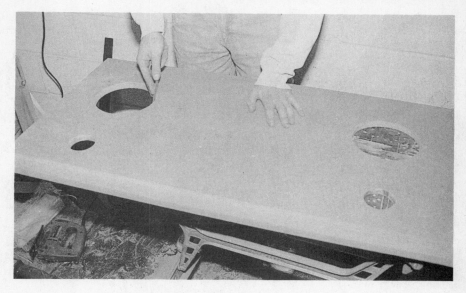

The latter should be laid flat on the shelf and, with the shelf inverted, the excess cut away.

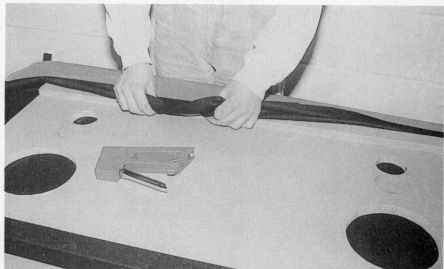

When it's been tacked into place, you can use a sharp craft knife to give it a final trim before—

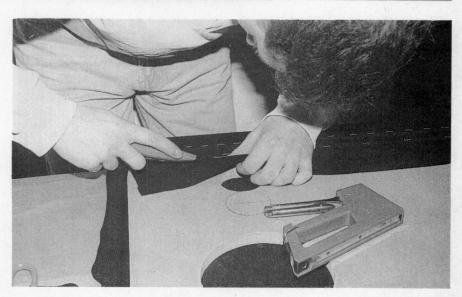

Speakers and crossovers

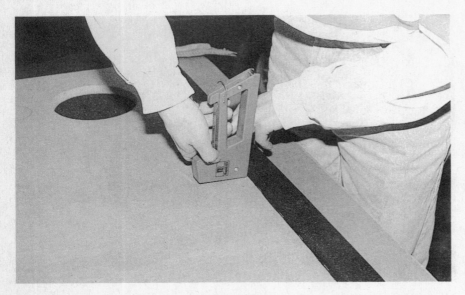

—-stapling the cloth in such a way that there are no loose ends or creases which could fray or catch as the shelf is raised and lowered.

You can use either strong glue or a staple gun to secure the edges of the cloth. Make sure that it is pulled very tight and is not crumpled underneath, out of sight. Keep checking and don't rush it.

Mounting the speakers is not difficult. Both speakers come with grilles for surface mounting, the mid-woofer being a simple clip- on/off affair, but the tweeter requires some slight dismantling before the grille can be removed. It would be possible to surface mount the PRS 602' on this shelf, but thieves have sharp eyes and it pays to keep them guessing as to what tasty equipment you have fitted.

The mid-woofer is secured by long self-tapping screws. Note that the speaker terminals are facing forward so that there is less risk of accidentally damaging them.

Speakers and crossovers

The hole for the tweeter was cut so that it was an exact push-fit. In order to make a good, sound-proof seal, silicone sealant was used around the edges.

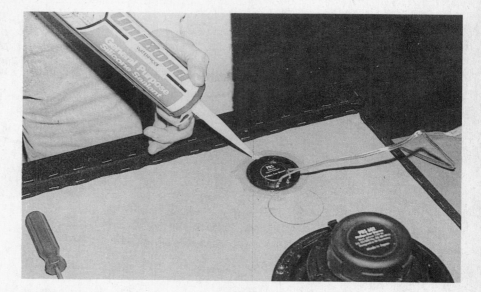

Even though the wiring would not be seen, great care was taken to keep it looking neat. Excess wire was trimmed and the wiring was stapled into position

The crossover terminals are simple screw-in type - take care to get the wiring right, however, or you'll get an unpleasant surprise when you switch on!

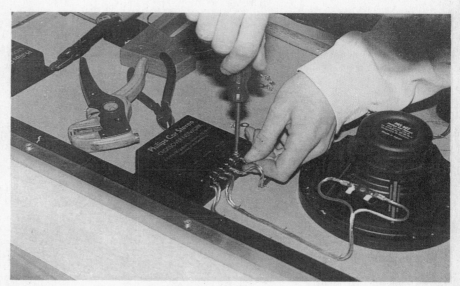

Speakers and crossovers

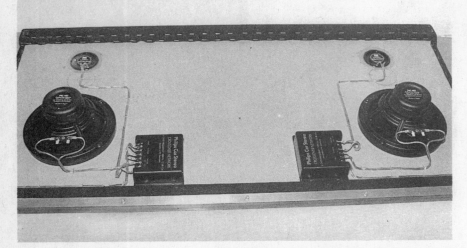

The finished shelf is an excellent piece of ergonomics as well as a practical and substantial improvement in speaker performance and when in situ—

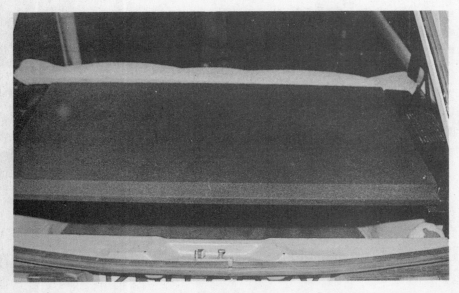

—there's little to tell the outside world that it sounds as good as it looks. The Dards fitter made it look easy and finished the job much quicker than most of us would, but there is nothing here that could not be accomplished by most of use, given a little time and patience.

One of the major problems with any rear shelf speakers in a hatchback is that folding nature of the rear seat makes life difficult when it comes to wiring. We solved this by routeing the wire from the set to the original speaker positions (at either side of the shelf) and then fitting a plug and socket arrangement to it and the wiring from the crossovers. This meant that, by unplugging them, the rear shelf could be removed altogether in seconds - an excellent measure of practicality.

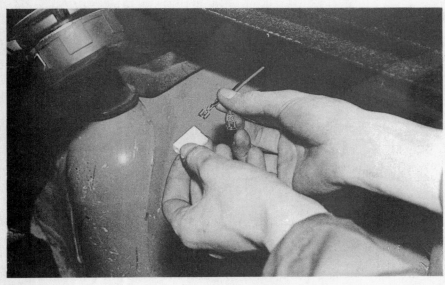

**TECHNICAL DESCRIPTION
Philips PRS602 Component speakers**
6" woofer & tweeter with 25mm titanium dome. 150W max power handling capability, frequency response 34Hz – 25,000Hz

Speakers and crossovers

SUB-WOOFER SPEAKERS AND SUB-BASS TUBES

If a woofer handles bass frequencies, it follows that a sub-woofer handles sub-bass frequencies, those below 200 Hz. These frequencies are felt rather than heard, a factor which means two things; firstly, there is no stereo effect, so there is no real need to have sub- woofers in pairs and secondly, there are few limitations on mounting position. The main qualification to that last comment is that they must be held securely, otherwise, all you will hear is a mish-mash of sub-bass waffle with no real clarity; bass sound rather than bass notes.

SPLITTING THE SOUND

The sub-bass frequencies have to be separated from all the other frequencies emanating from your sound source. This means that, somewhere down the line, a crossover unit is required. Once again, there is no one, simple answer. It depends on your requirements and those of your system. If you've read the section dealing specifically with crossover units, you'll know that a passive crossover splits the sound after it has been amplified and an active crossover (better, but more expensive) sorts out the frequencies before amplification. Some amplifiers have built-in crossovers, enabling you to run a sub-woofer direct. Many of the good quality graphic equalisers have a line-out facility for sub-bass sound, again, making life easier.

AMPLIFYING THE SOUND

Like all speakers, the sound going into a sub-woofer speaker has to be amplified first. It is common to bridge stereo outputs of an amplifier in order to power a single sub-bass speaker. Because of the nature of the sound, you will need a disproportionately large amp compared with the rest of the system. For example, it is not unusual to power a large single sub-woofer speaker with 100W, when the four speakers in the cabin only have 100W between them. This is because a lot of air has to be shifted in order to reproduce the low notes, and this shifting requires a lot of power.

Ideally, a sub-bass speaker should be housed in a purpose-made cabinet, designed to suit the characteristics of the speaker and its working environment. However, this is highly complex and skilled work and beyond the reach of this book. It is possible, though, to buy ready-made cabinets suitable for certain sized speakers. Remember that they must be well secured in the car (for safety as well as sound quality) and that they will take up no small amount of luggage space.

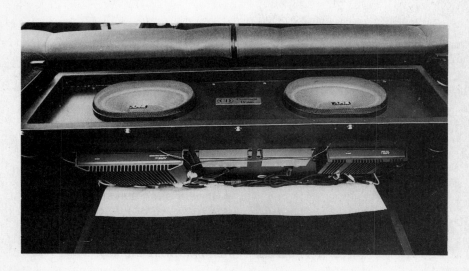

The mounting of speakers is extremely important; the larger the speaker, the more secure it has to be and sub-woofer speakers have to be the most secure of all. A speaker which is rattling around is little more than useless. Here, some hatchback space has been sacrificed in order to make a secure mounting for the twin sub-woofers and also to make a cabinet which effectively forces the sound into the car, it can't go anywhere else. The strong plywood construction makes an ideal baffle to prevent the low frequencies shaking your fillings free, although many installers favour MDF (medium density fibreboard). The other advantage of such an installation is that it provides a suitable mounting place for the amplifiers and crossover units which are part and parcel of an extensive system such as this.

Speakers and crossovers

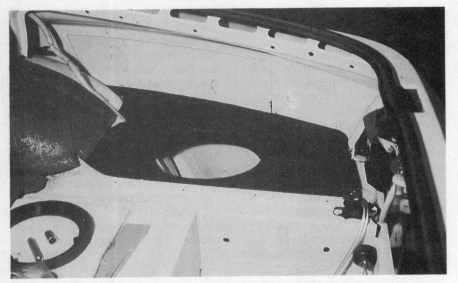

Some cars lend themselves to sub-woofer installation. Here, a widened rear wheel arch creates a natural cabinet which only needs a top baffle. In this case, it was created from plywood and a 6" x 9" sub-woofer speaker screwed firmly down into a suitably accurate hole. In order to ensure that the frequencies did not escape, the top of the 'cabinet' was sealed with silicone sealant and the wheel arch was lined with sound-deadening material.

You can go to extremes! This is Dards Electronics BMW show car, designed to enter Sound-Off competitions and thus not a run-of-the-mill installation. The huge Philips sub-woofers are mounted in a specifically designed cabinet which can easily be removed—

—should luggage space be required. Note the hole in between the two amplifiers - through this the sound is ported along the underside of the car up to the driver and front passenger. The system uses only two amplifiers, mounted either side of this port. When you hear such a system, you will realise that you don't have to fill the car amplifiers to excel – correct and thoughtful installation is the key.

Speakers and crossovers

SUB-BASS TUBES
One of the easiest (and not surprisingly, increasingly popular) methods of getting that low-down bass sound is to use a sub-bass tube.

The most obvious drawback is that, when installed in the normal manner, it cannot be held as firmly as a purpose-built sub-woofer box. This leads their detractors to say that it can never produce the clear bass sound in the same way as a fixed speaker. Doubtless, the argument as to whether speaker or tube is best will rage on, but for many it boils down to pure practicality - if you haven't got room for a sub-woofer speaker but you want sub-bass sound, then you've got to have a tube!

Some tubes have speakers in either end and some have internal amplification (which increases the cost, but saves on wiring and mounting problems). One of the main advantages of a tube is that it only takes-up luggage space when you feel like it; if you have a need to carry lots of luggage, you can unplug it and run your system on just the cabin speaker system. It is this versatility which has seen their popularity rocket in a very short time indeed.

A second advantage is that the enclosure (the tube itself) is designed to be exactly right for the size of sub-bass speaker and its handling capacity, thus compensating (at least to some extent) for the relative lack of stability.

It has to be said that, unless you have a particularly good ear, or you are comparing extremely well- qualified (ie, expensive!) systems, it is very hard to tell whether the sub-bass is tubular or 'normal'.

WHICH TUBE FOR YOU?
Like most other equipment, this is a matter of personal taste and, of course, the depth of your pocket. At the time of writing, it is hard to find a poor sounding tube. Doubtless, as their popularity brings more 'spurious' manufacturers onto the market, then it will become more important to choose carefully. The best advice is to listen to a number of tubes. Whilst you can do this at an ICE specialist's showroom, it is better to find a dealer who is running a demonstrator vehicle with a tube installed. After all, it is meant to perform in a mobile environment.

What you should be looking for is a hard 'thump in the back' type of bass sound, rather than a 'woofly' mumbling, noise from the lower registers. Again, comparison is the only way to judge which is which. Another factor is your personal taste in music, heavy rock music will make different demands on a tube from, say, country and western. Again, it's a question to put to the experts, note the use of the plural there. Whatever you buy, try to get more than one opinion. Consider also the physical size of the tube, both diameter and length. At present, 8" diameter is the most popular, with 6" and 10" selling in roughly the same proportions.

Bazooka popularised the tube in the UK and now offer a wide range of sizes and types.

Speakers and crossovers

In common with many ICE equipment manufacturers, Philips market their own tube, this model being the Sub-bass tube 150 shown here in their Mitsubishi Shogun demonstration vehicle.

FITTING A SUB-BASS TUBE

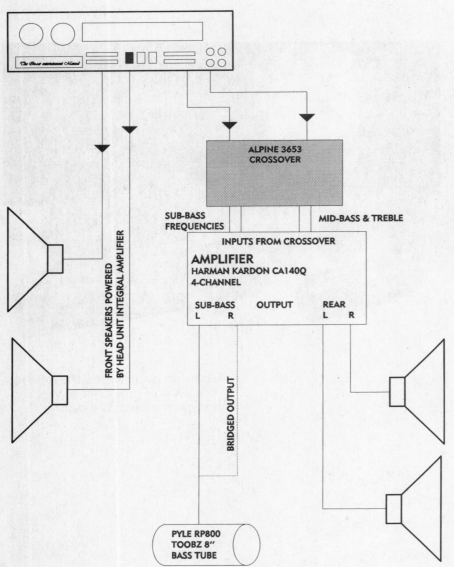

In this section, a Pyle RP800 Toobz sub-bass tube was fitted to a Ford Orion by Dards Electronics. It was already equipped with a high quality radio cassette deck with an integral amplifier which was being used to power the four co-axial speakers in the cabin. However...

... it also had a line-out facility, so, the deck was used to power the front speakers, whilst the two rear line-out cables were taken directly to a Harman Kardon, CA140Q amplifier. This is...

Speakers and crossovers

... a high-power, 4-channel unit, producing 4 x 35W or 2 35W and 1 x 70W (bridged). The latter combination was ideal for our purposes and we took two of the channels to the rear speakers and the remaining two to the sub-bass tube.

As is usual, the connections are made on one end of the amplifier, although a neat feature of the Harman Kardon unit is that they are recessed into a plastic moulding. The connections are thus protected from damage by errant feet or luggage. The usual care and attention should be taken when wiring the amplifier, as shown in chapter 11. In order to separate the frequencies...

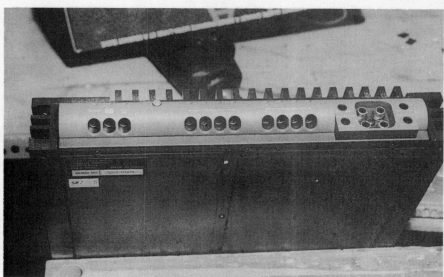

... we fitted an Alpine 3653 active crossover. Note the markings on the top of the unit, which are a graphic illustration of how the crossover points can be varied.

Speakers and crossovers

The Pyle tube itself is of American origin (you may have guessed from the name!) and offers an excellent bass sound, particularly when referenced to the price. Another important point is that the speaker in the end of the tube is superbly protected by a thick metal mesh - some tubes only have the equivalent of chicken wire!

Don't forget that all these items need linking together. In the Dards Workshops, the right wiring and connectors were always on hand, but you will need to think your fitment out carefully so that you have enough cabling etc. to finish the job. It's wise to use pre-wired RCA connectors etc. wherever possible.

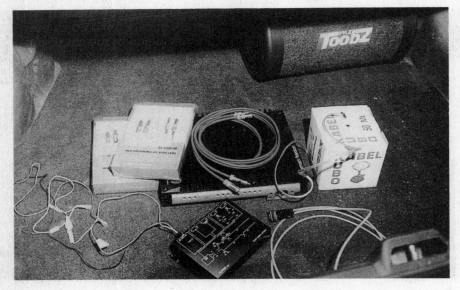

The wiring instructions with the Harman amplifier and Alpine crossover are explicit and simple to follow as long as you take your time. There are plenty of connections to make and it would be very easy to get it wrong and damage lots of expensive equipment. There are a number of amplifier diagrams, showing the various permutations available. The one shown here is the one we used. (Courtesy Harman Audio Ltd)

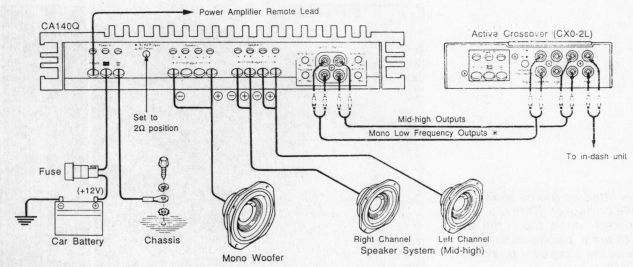

Note: * If only one mono jack is provided, a "Y" connector can be used.

Speakers and crossovers

The amplifier and crossover were mounted on the back of the folding rear seat. It would have been possible to fit them under the front passenger seat, though this would have left them more open to damage from rear seat passengers and it would also have meant more wiring difficulties. Tim made sure that both units were securely fixed and...

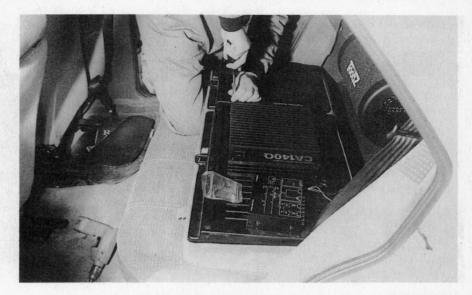

... took care to loom wires of like kind (ie not speaker and power carrying wires) together. This not only looks neater, it's safer, too. Note that the amplifier is mounted in such a way to allow the heat generated to escape along its fins, i.e., they are vertical rather than horizontal.

The tube wiring couldn't be simpler, just two wires to clip onto the terminals. Note the hole above, which is a port, designed to allow the sound from the tube to 'breathe more easily'.

Speakers and crossovers

An important professional tip from Tim was to leave a small loop of wire at the terminals and then secure the cable to the tube. This way, if the end of the cable is jerked for any reason, the connections will not be broken.

The Toobz is secured by means of an adjustable webbing which is secured in the car by means of these plastic clips which are self-tapped into place. If extra luggage space is required, the straps ensure that the tube can be released in seconds.

The finished set-up is extremely neat and it sounds good, too. The tube position is governed largely by the needs of this owner to make almost constant use of the luggage space available. The sound a tube produces can vary quite considerably with its position in the car. Some, for example, work better when they are 'fired' into the corner of a vehicle. Clearly, this is a case for some experimentation, though, as in this case, practicalities may leave you little or no choice as to where to put the tube.

TECHNICAL SPECIFICATION
Pyle Toobz RP800
8" sub-bass tube, 200W max power handling, black carpeting exterior finish, with rear port. Length 450mm, height 275mm, width 275mm, weight 5.5 kgs.

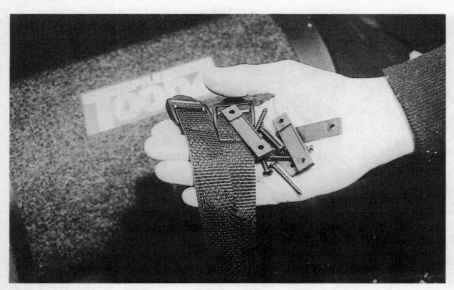

Harman Kardon CA 140Q Amplifier
4-CHANNELS, 4 x 35W (o.1% THD) or 2 x 35W (O.1% THD) & 1 x 70W (0.3% THD)
S/N ratio; 100 dB
Frequency response: 5 Hz to 100kHz

Alpine 3653 Active crossover
2-way dividing network, 24K gold plated RCA input/output connectors. Continuously adjustable crossover frequency. Stereo/mono switch for sub-woofer systems.

Speakers and crossovers

FAULT DIAGNOSIS

No sound from speakers, or intermittent sound	Make sure that the set is switched on and that there is a sound source radio/tape etc.). Check terminals on back of speakers and connections at back of set. Check routeing of cables to ensure that they are not trapped. If you are running separate power amplifiers/crossovers, make sure that the plug-in connections are secure.
Unbalanced sound	Check fader/balance controls on set. Check that speakers are connected correctly (right/right, left/left).
Phasing	Caused by incorrect polarity. Check all speaker connections and ensure that positive and negative connections are followed through the system.
Speakers rattle when volume turned up	Isolate problem to specific area. Secure speakers and grilles. Check security of trim panels. Extra baffling may be required.
Front door speakers stop working after a few days	Check speaker connections at the set. Ensure that there is enough slack in the cables where the wires leave the doors. Check terminals on the back of speakers.
Crackling or extraneous noise	Isolate the problem; is it all speakers? Swap the leads around; if the problem sound 'moves' then the fault lies with the set. If the problem sound stays, the it is the speaker at fault. Does the problem occur solely on radio broadcasts? Radio interference should be present on all speakers. (See chapter 7) Hum from some of the speakers in a system indicates a possible wire routeing problem, e.g. running a speaker lead too close to a power-carrying wire.
Vibrations from speaker or surrounding trim	Speakers have to be mounted securely to perform at their best. Bass frequencies will soon show up a poorly mounted speaker. Check that the speaker is held firm. You may need to use an extra baffle. Wherever possible, don't fix speakers in panels solely to the trim and if you do, use the plastic securing rings supplied with most speakers

Chapter 4
Cassettes, description care & storage

You don't have to look very hard to see that there are many varieties of cassette tape on the market. Although there are many manufacturers, there are just three types of cassette; *types* I, II and III. As can be seen from the table here, the better the quality of tape, the higher the bias figure and the lower the EQ (equalisation) figure.

Whilst the highest quality tape is metal, it should be noted that this can only be used on tape decks which are equipped with suitable heads. Using metal tapes in decks without these heads can ruin the heads altogether, a very expensive business.

Other factors which come into account in tape reproduction (apart, of course, from the quality of the deck itself) include the manner in which it is recorded. Clearly, the best tape in the world will not produce a good sound if it is recorded on a poor quality machine. There is also machine preference, an intangible which defies rhyme or reason and which you may never even encounter. It could be that brand 'A' tape performs well on your equipment and yet an equally high quality brand 'B' tape sounds awful. The answer is to find the tape which performs best by a process of elimination. When it comes to the brand of tape, the recommendation has to be to buy from well-respected suppliers (TDK, Sony etc.) You will usually pay a little more for your cassettes this way, but the performance will reward your investment. It is sometimes possible to get better results from a Type I cassette from a quality manufacturer than a Type II cassette from a 'spurious' supplier.

Cheaper tapes tend to deteriorate after a while, usually by shedding their magnetic coating all over your cassette heads, landing you with a double problem. The resultant 'recorded in a broom cupboard' sound should be enough to put you off cheap tapes for life. Most domestic cassette players have switchable bias and EQ and thus, can be altered to suit the various types of tape. Few in-car units have this facility, although some top of the range sets will automatically switch to metal mode (which usually includes chrome) where appropriate.

TAPE TYPE	ALTERNATIVE BIAS NAMES	EQ	
Normal	Low noise/type I	100%	120pu sec
High bias	Chrome/chrome position/ Type II	130%	70pu sec
	Metal Type III	200%	70pu sec

Cassettes, description, care and storage

Normal Bias Tape
This is composed mainly of gamma haematite which is a variety of ferric oxide. This can be used quite safely with all cassette equipment.

High Bias Tape
High bias, or Chrome (Cr02) tapes, have come to the fore in recent years. These tapes can utilise higher bias and EQ to produce a much better sound from a wider frequency response.

Metal Tape
Metal tape is a highly specified variety of tape which has, for various technical reasons, a very high resistance to magnetisation.

This means that, when recording, a much stronger signal can be stored on the tape and this strong signal holds down residual noise and avoids distortion.

As well as the quality of the tape you should also consider the length. Never go higher than a C90 (45 mins each side). The C120 (60 mins each side) is forbidden by all in-car tape deck manufacturers and me! In order to get that much tape physically inside the cassette, the tape is made very thin. Before long, it starts to slip on the spools and eventually it will jam. Untangling 20 metres of tape from the delicate innards of a hard-to-get-at deck is no fun.

Tapes can be purchased at almost any length, even down to five minutes per side. Ultimately, you can customise your tapes to suit by using a cassette tape splicer, as shown in this chapter.

WHICH TAPE TO CHOOSE
If you have a basic model cassette deck with no frills or fripperys, there's an obvious temptation to use the cheapest cassettes you can find. But if you trade up in the future, you'll find yourself with a load of cassettes which sounded OK on your old set, but now sound pretty poor. Buying metal tapes is good policy if you have the equipment on which to play them. However, bear in mind that they are the most expensive tapes to buy and that if you want to play them on any other deck, then that too will have to be metal compatible. A good, centre ground recommendation is to buy Type II tapes from a reputable manufacturer. Pricewise, they fall in-between the Types I and II and the quality is close to that of metal, particularly in the noisy in-car environment. In terms of sound quality, they will give good reproduction on most sets.

Remove the top gently, so as not to disturb anything, and take note where everything goes, in particular, how the tape winds around the spindles. It's a good idea to draw a diagram which will help when it comes to reassembly. Always avoid touching the tape wherever possible and keep it away from sources of magnetism.

Some cassettes are glued together, but the better quality ones are constructed using small, cross head screws. These are easily removed using a suitable screwdriver. Note that they can easily be damaged by using the wrong size screwdriver! There are usually five screws in all.

Cassettes, description, care and storage

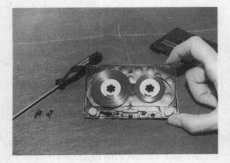

The innermost workings of a standard, C90, cassette. There are lots of delicate bits and pieces and all must be handled with care, particularly the tape itself, which is incredibly thin.

Eventually, no matter how careful you are, no matter how good the tape you buy, the inevitable will happen; the tape will break. Often this is as a result of a machine problem which causes the tape to wrap itself around the pinch rollers, capstan and just about anything else inside the deck. To retrieve it, remove the deck from the car if possible, if not, disconnect the power source, and use something non-metallic to fish-out the errant tape. Be extremely careful and patient as cassette tape is very thin and damages easily. Try to avoid putting your fingers on the tape surface. If you have a tape splicer, you can save your valuable tape (and its contents) quite simply. Naturally, you will be able to hear the join, albeit only for a fraction of a second. If your tape has been badly damaged, then you may have to cut some out. However, it's still much better than throwing the whole lot away. Don't try to use tape which has been badly mangled, as it will tend to jam once more, leaving you back at square one.

Although most of this section deals with 'normal' cassette tapes, there is a newcomer in the fold. Again, a Philips invention, DCC (Digital compact cassette) is set to rival compact disc in terms of sound quality.

CASSETTE CARE & STORAGE

Having obtained the correct type of cassette for your requirements, it is up to you to look after it. Always keep cassettes out of direct heat (sunlight, car heaters etc.) and extreme cold. Store the cassettes in their library cases or, better still, in a cassette carrier which holds the spindles secure and prevents the tape itself from moving around when not in use. Always keep tapes away from sources of magnetism, as this could seriously impair the sound quality. As the transportation of tapes in a car is the biggest cause of damage, this section is devoted to ways of overcoming the problem. When mounting any form of in-car cassette holder, you should consider carefully the safety aspects. Holders should be of good quality and mounted very securely so that they cannot break loose and fly around the cabin like an unguided missile in the event of an accident.

Cassettes, description, care and storage

Fischer C-Box make excellent quality products. All their cassette boxes are robustly finished, using safe, shatterproof materials. They are stylish, and designed to blend in with the interiors of most cars. Shown here is a small sample of their 'tailored to fit' range, for which they are justly famous.

This photo shows how easy it is to keep your cassettes safe and secure in the BMW 3-series. At top, the standard armrest is seen in the 'down' position, revealing nothing of any interest to a casual thief. Lift the lid, however, and there is storage space for six cassettes.

Cassettes, description, care and storage

Utilising the centre console space which usually ends up cluttered with odds and ends is a Fischer speciality. This Escort benefits greatly by the addition of this tailored-to-fit, 6-cassette C-Box

For those who own cars where it is not possible to fit a purpose-made C-Box (or who have already used the space) Fischer produce a range of universal units. This is the 4-cassette universal carrier, which mounts either vertically or horizontally on the bracket provided. As with all Fischer products, this box holds the cassettes without their cases and in such a way that the spindles are prevented from turning.

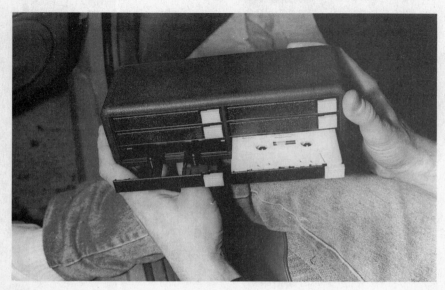

This is the 6-cassette universal model, which can be mounted—

Cassettes, description, care and storage

—in any one of a number of ways using the bracket provided.

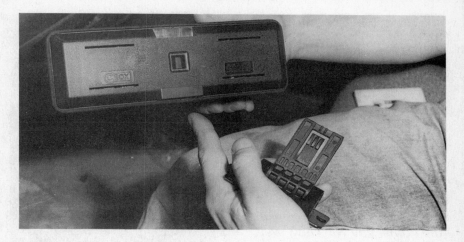

This can be stuck or self-tapped wherever it is required. On the dash top, as seen here, is often the only place left for cassette storage. However, you should be aware that it is a place where the sun will heat the box and its contents and it will be immediately obvious to the prying eyes of a thief. Also, it should be fastened very securely, least it become a danger in the event of an accident.

It could also be mounted under the dash; in the centre, as here, is best, so that both driver and passenger can reach it.

Cassettes, description, care and storage

Like many C-Boxes, the portable has a novel method of showing whether each tray is 'occupied'. As the cassette is inserted, a red plastic tab is pushed down in front of a perspex bubble, thus giving a clear, visual indication.

If you use cassettes in the car but use the same cassettes in the home, you will know well the irritation caused by wanting the listen to a tape and finding that it is in the house/car! One answer is this Fischer portable box. It holds 12 cassettes without their cases and, whilst it can be carried loose in the car, brackets are provided so that it can be secured in position (though easily removed). With an item like this, you are well advised to take it with you every time you leave the car,- a box full of cassettes (and the C-Box itself!) is well worth breaking a window for.

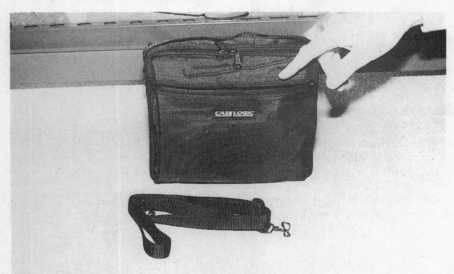

As an alternative to the Fischer method of carrying tapes around, Caselogic have number of cases available. This one is the mid-range model, capable of holding 20 cassettes in their cases. The inner plastic tray keeps them in order and adds some protection against accidental damage and the nylon case is both durable and stylish. I like black, but you can opt for more outgoing colours if you prefer.

It comes with a detachable should strap and a handy pocket, useful for storing oddments, headphones etc.

Cassettes, description, care and storage

TAPE HEAD CLEANING

A much neglected task, but necessary if you want to get the best from your deck and tapes. As you play your cassettes, a thin film of magnetic particles builds up on the tape head. It is this that you have to remove. A cheap, though rather awkward method is to use a cotton bud soaked in solvent. This should be poked into the cassette aperture (no cassette in situ, of course) and used to clean the heads and rollers. It is a little tricky on an in-car unit, even if you remove the deck from the car. Without doubt, using a proprietary head cleaner is much easier and, in most cases, far more efficient.

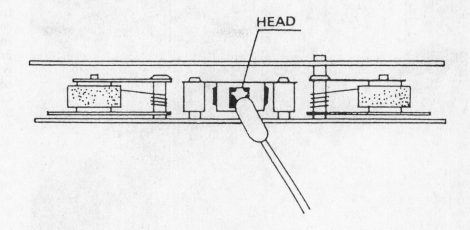

One of the easiest and most effective ways to keep your tape heads clean is to use the Allsop 3 system. There are various permutations of the company's products, but for the enthusiast, this is the best and most complete. In this handy and strong plastic case, there is a cleaning cassette, a bottle of cleaning fluid and a number of spare cleaning heads. It takes up little space and could easily be carried in a glove box or map pocket.

Using it is simple. The ozone-friendly cleaning fluid is applied to all five cleaning heads on the special cassette which is then played in your deck for 30 seconds - each way, on an auto-reverse unit.

Cassettes, description, care and storage

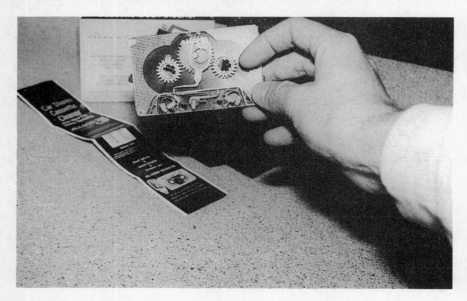

The cogs in the cassette push the cleaning heads rhythmically onto the tape heads, cleaning as they go. A both simple and thorough solution.

HINTS AND TIPS

1. Always keep the tape in the cassette mechanism taut. Winding the spindles manually with a pencil is a simple way to do this. If the tape is left to become slack, then the tape could easily spill out into the machine and snarl around the pinch rollers.

2. Always make sure that the magnetic playing head, the pinch rollers and the capstan of the machine are kept clean, using one of the methods shown earlier in this chapter. In addition, the head should be de-magnetised on a regular basis.

3. Always store the cassettes in their library cases or purpose-made car cassette carrier. Either method ensures that the tape spindles are held in position and will not allow the tape to become slack. Of the two, a carrier is to be preferred, as, stored individually, there is a tendency for tapes to end up in all sorts of places (under the seats, on top of a hot dash) not recommended for such precision items.

4. Keep cassettes out of direct sunlight as this can cause tape performance to deteriorate and/or the casing to warp.

5. Avoid touching the tape itself with your fingers, the result of doing so usually results in 'drop-out' or unwanted noise.

6. Always keep your cassettes away from any form of magnetism.

7. Leaving your cassette in the deck whilst not in use is not recommended. Some machines have a key-off mechanism which releases the tape inside the machine and prevents it (and the machinery) from being stressed for long periods of time. Even so, the best place for a cassette not in use is in its library case of a purpose-made carrier.

Chapter 5
Choosing and fitting a Radio Cassette Deck

Even a brief look into any high street audio showroom will tell you that there is plenty of choice when it comes to in-car radio/cassette decks. A first, vital point to remember is that the deck is part of a system, and even the simplest stereo system will have at least two speakers, and it is speakers, regardless of number, you must consider first. It's no use buying a super-powerful deck only to play it through your old speakers with a power handling capacity barely half that required. It won't sound very good, to say the least, which means that you'll have wasted your money and you'll have to buy some aspirins to cure your headache!

WHERE TO SHOP
Not so many years ago, the specialist ICE dealer was someone to be consulted only by the out-and-out enthusiasts with deep pockets. Not any more; the recessionary times of the late 1980s had the effect of sorting out the wheat from the chaff and the result is that there are a large number of such dealers who can offer some great deals either on straight DIY sales or installation as well. The clear advantage of a specialist is that you are likely to be able to get some well qualified advice, something not always available in the supermarket-type environment. Given the massive number and variety of sets on sale today, you can hardly be expected to know everything about every set and so asking a few pointed questions is going to be necessary. The answers you receive (or not) should help you make up your mind either to buy a specific set or to try another retail outlet! With most of us, the key factor is cash. In your search to find the most features for the least money, you will probably find that you have to pay a premium for a 'name' and that, often, well-known brands appear to offer less in real terms. However, established names do not become so overnight and a good reputation takes some earning. Consider also the service and back-up spares facilities, not only of the manufacturer but also of the dealer in question. Your bargain purchase may no look so rosy if you find yourself waiting six weeks for a 'simple' repair. You won't be happy, and I know, because it happened to me some years ago!

READ ALL ABOUT IT
There are several specialist ICE magazines where experts test machines and give their expert opinions. These are a good starting place as they will often directly compare sets and print the results of straight 'fight'. In addition, they list a large number of sets in a glossary which lists their major features together with their price. Some mags are better than others in this respect. For impartial and

Choosing and fitting a radio/cassette deck

informed comment, you can't do better. However, tests of decks should be taken only as a basis for narrowing down your choice. Remember that the element of personal preference has to be present in any test and that what looks like a large difference in technical specification (when the equipment is linked to lab equipment) may be irrelevant once that deck is installed in a vehicle. Consider also how criticisms of machines apply to you. For example, a 10-disc autochanger may be criticised in a magazine for being bulky and awkward to install, but if your vehicle is a large luxury car or even a van, finding space to install it will be the last of your problems

WHAT TO LOOK FOR

In terms of user features, the car audiophile has less to worry about than his domestic counterpart. On domestic equipment, many of the features offered are simply there as 'bait'; gimmicks to make a machine appear more attractive than a competitor, when there really is not much difference.

However, on an in-car machine, most features are not only functional but very necessary, designed to keep the driver's attention focused firmly on the job in hand, safe driving. Anything 'automatic' or 'programmable' is good news. Don't be afraid of apparent complexity. Even the most complex of machines are quite user friendly nowadays, and a short while spent with the handbook should soon have you totally *au fait* with your new purchase.

When you browse through the technical description of any radio/cassette deck, you will find a whole host of jargon-based terms. These will be very useful to you, as long as you understand what they mean!

Peruse the glossary at the back of this book to get a basic idea of what's what.

A cassette tape noise reduction feature is a good start (see below) and auto-reverse is both convenient and safe. When the cassette reaches the end of one side of the tape, it automatically starts to play the other. This also means that you will have the facility to swap from one side of the tape to the other at the press of a button, rather than having to remove the tape from the deck. If you intend to use metal tapes, then your deck **must** have metal tape facility, otherwise you could seriously damage your heads. Some decks have a facility for checking the type of tape being used and adjusting the settings accordingly. Others require you to set them manually.

Are you going for two speakers or four? Whilst you can adapt a two speaker set to four, it's extra work and it does mean that you'll have to add a front/rear fader if you want to have full control over your sound balance.

NOISE REDUCTION

Excess noise has been a problem with the compact cassette ever since its inception. The problematic noise takes the form of playback hiss which is, in fact, electromagnetic noise generated by the tape (which moves quite slowly) coming into contact with the playback head. (As an aside, this is one of the reasons that DAT provides such good quality sound - it moves more quickly across the heads.)

The 'hiss' is most prominent when a quiet piece of music is being played. Clearly, if a recording is made with high input passages (heavy metal, for example) then tape hiss will hardly be noticeable, as the recorded sound will be much louder than the hiss. Some form of noise reduction is desirable on any cassette deck and becomes increasingly so, the greater the power output, whether from an integral amplifier or from external power amplifiers.

DOLBY 'B'

There are many forms of cassette tape noise reduction system, but without a doubt, the best known and widest used is the Dolby 'B' system, developed by Ray Dolby. It is designed to boost the signal-noise ratio by around 10 dB in the frequency range about 5 kHz. Although other systems may offer improved performance, Dolby 'B' scores heavily by being compatible with a vast range of equipment and most commercially available prerecorded tapes use Dolby

DOLBY 'C'

This works on a similar basis to Dolby 'B', but spans a wider frequency range and gives another 10 dB (a total of 20 dB) of noise reduction. In addition, it is so designed to play back through a deck which is equipped only with the early format Dolby.

DBX

DBX is not strictly a noise reduction system, although it does have that effect. It is a massively popular dynamic range expansion system, boosting as high as 100 dB in certain cases, with little sign of tape hiss. In the process of recording with DBX, the signals are radically affected and, as such, DBX tapes should not be played on sets without the relevant DBX circuitry.

SUMMARY

Tapes recorded with no noise reduction encoded on them, should be played back that way. Not doing so will result in a dull, muffled sound. Playing a tape encoded with Dolby 'B' with the noise reduction feature 'off' will

Choosing and fitting a radio/cassette deck

result in the sound being very sharp and sibilant. Repeating this operation with a Dolby 'C' encoded tape will sound even more 'trebly'. Playing a Dolby 'C' encoded tape through a deck with Dolby 'B' switched on will sound good, though not as good as it would if played through a deck with 'C' available.

DBX encoded tapes should not be played back at all unless the set has DBX facility switched on.

FREQUENCY RESPONSE

Frequency response (FR) the denotation of the highest and lowest note a deck/tuner is capable of reproducing. It is given in Hz or kHz, with the lower figure being the bass end of the scale and the higher figure being the treble.

POWER OUTPUT

Most radio/cassette decks have a built-in (integral) amplifier. Power output is the factor that most people are aware of and that, unfortunately, many people give too much importance. There are two basic methods of defining power - Maximum (max) and Rated or continuous (RMS). The maximum power is that which the amplifier within the set can produce for a short (sometimes very short) period of time. The rated power is that which the amplifier can produce on a continuous basis. The latter will be the lower figure and the smaller the difference between the two figures, the better the amplifier. Check also the THD (Total harmonic distortion) figure for the quoted outputs. When comparing sets, you should ensure that you are comparing like with like, difficult though it may be. See chapter 11 for further details about what to look for in an amplifier.

SPEAKERS

Speakers are dealt with in detail in chapter 3, and you should cross-reference accordingly. Most radio/cassette decks come without speakers, on the assumption that you already have a pair/set in your car. The important points to check are; the power handling capacity, the frequency response and the impedance. In general, an uprated deck will demand uprated speakers, especially if you are running those fitted as standard by the vehicle manufacturer.

SECURITY

Some form of security feature is welcome on any radio/cassette deck. Quick-out, whereby you take the set with you is offered by most companies, though it is becoming common for the front control panel, rather than the whole deck to be removable. Coded decks are ones which require the programming of a special code number before the set will work after the power has been disconnected. Some sets have electronic locks which secure the deck physically in the dashboard when the ignition key has been removed. Some Roadstar decks feature a built-in siren which sounds for 30 minutes, should the deck be forcibly removed from the car! Any security feature is better than nothing, though the key factor is to be aware that all the time, someone, somewhere, is waiting to steal your stereo equipment - stay awake! (See chapter 13, security)

UPRATING YOUR SYSTEM

If you are buying a deck to form the basis of an uprated system sometime in the future, then you've got to buy quite a good set now. Some form of noise reduction is required for, if you add some heavy amplification to a set not so equipped, the tape hiss will be louder, along with the music. A set with at least two 'line-out' sockets will make adding amplification easier - you simply plug-in. A deck with four line-out sockets and no integral amplifier is usually referred to as a pre-amp set. These are usually very well qualified sets, toward the top of the price range and, of course, require at least one amplifier before they can be used.

Consider the addition of a CD autochanger. Many machines (such as the Clarion CRX 91R shown later in this chapter) have the plug-in capability which makes life much easier if you are going digital later on.

TUNER BASICS

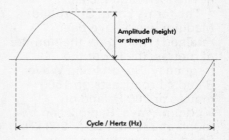

Radio waves are electromagnetic and hurtle through the air and 300,000,000 metres per second. There are several parts to each wave, as can be seen in this diagram. The cycle is the width of the wave and the amplitude is the height or strength of the wave. The number of times a wave goes up and down in one second is called the frequency and this is measured in Hertz (abbreviated to Hz). For example, if a cycle takes one second, then it has a frequency of 1 Hz. Broadcast frequencies are much higher than this and are denoted in either thousands (kHz) or millions (mHz) of Hertz.

WAVEBANDS

The Wavelength is the distance of one complete wave, measured in metres. Amplitude is the strength or height of the wave. Radio stations send out a 'carrier' wave. For Radio 2, for example, this is 909 kHz and

Choosing and fitting a radio/cassette deck

when the radio tuner is tuned to this frequency, all others are ignored. Acoustic signals, either music or speech, are then sent along this carrier wave. As the sounds vary in pitch or volume, so the amplitude of the carrier wave varies. The variation is called modulation. The process is called Amplitude Modulation, or AM for short. The alternative method of modulating the carrier wave is to vary its frequency rather than its amplitude. Not surprisingly, this is called Frequency Modulation or FM. Radio sets used to be calibrated in metres, but most now show frequencies.

The terms Long, Short and Medium wave have been retained. The higher the number of metres, the lower the frequency. So, on Long Wave, Radio 4 broadcasts on 1500 metres or 200 Hz and on Medium Wave, Radio 2 can be found at 330 metres or 909 kHz. Short wave frequencies are much higher, although these are not widely used in automotive applications.

Short waves do not travel far at all but by their very nature, are reflective and can effectively 'bounce' over long distances. However, this means that some areas can receive no SW transmissions at all.

Long waves, as the name suggests, are capable of travelling long distances, often hundreds of miles. However, they are very susceptible to interference.

Medium waves do not travel as far and are particularly prone to interference in the evening or at night due to atmospheric conditions. They have a limited frequency response, especially when compared with FM. The greatest span available on AM is around 100 – 9,000 Hz, although often it is much less.

VHF is most popular with the sound enthusiast, largely because this waveband is (as yet) the only one broadcasting in stereo. In addition, FM stations transmit a wider frequency range, usually between 50 - 15,000 Hz. FM also has louder 'loud bits' and quieter 'quiet bits' as with live music. This wide dynamic range approaches that of Compact Disc. Finally, FM is less prone to hisses, crackles and pops than AM. One major drawback of FM is that the VHF waves that modulate on FM travel in 'sight lines' and can easily be blocked by tall buildings, tunnels or high ground. The range is not particularly impressive, even if there are no obstacles to overcome; a range of around 60 miles is the maximum for any one transmitter.

SENSITIVITY
This part of a radio tuner's specification is measured in microvolts (denoted as uV) and there will be different figures for FM, LW and MW. In techincal terms, it is the size of the voltage at the antenna (aerial) input required to obtain a noise-free sound signal. This means that the smaller the figure, the more sensitive the receiver is and the better the reception will be. To give some idea of what to look for, a top-range (expensive) deck should give the following figures;
FM   2 uv
MW   12 uV
LW   20 uV
Don't worry if you set doesn't match these figures; values of 5, 50 and 40 uV respectively, are quite acceptable for an average receiver.

TUNER FEATURES
Look for as many automatic features as you can get, prefer an electronic tuner to manual and FM is a necessity if you want to listen in stereo. A set with the facility to preset your favourite stations on easily used buttons is best for keeping your attention on the road. There's no need to look at the display in order to find another station. Auto-scan will, at the press of a button, allow you to listen to 10 secs of a station at which point, the tuner will pass on to the next. When you find a station you want to listen to, hit the 'scan' control again and that's it. This saves constantly having to hit the 'seek' button whilst driving. A handy feature is 'best station memory' (a function given different names by different manufacturers) whereby a single button will send the tuner to look for the 6 (the number varies) strongest stations and load them onto the presets for you.

This map shows the distribution of transmitting stations in the UK. The main stations are named and denoted by small circles, whilst the relay stations are marked by a triangle.

Choosing and fitting a radio/cassette deck

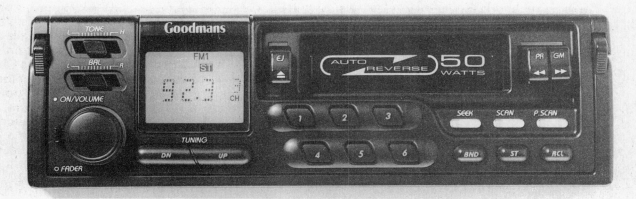

The choice of radio/cassette deck is staggering. A multitude of manufacturers each offering a long list of combination units can make life confusing for the unwary. In the popular budget end of the market, Goodmans are adept at getting as much value as possible into each deck. This GCE 232 is a 50W max set, with autoreverse, digital tuner with presets and quick-out facility. The styling is far from austere, something reflected in...

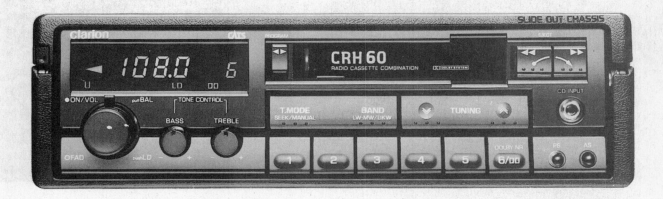

... Clarion's CRH 60, which echoes the tasty styling of the decks much higher up in their range. It's a similarly specified set, though with two security methods, quick out and CATS security coding. Note the inclusion of a CD input socket, suitable for a portable CD player, such as that shown in chapter 9.

Sharp is another company operating in the most competitive price range and this set is typical; 100W max output, in-built 5-band graphic equaliser, quick release and Dolby NR are just a few of the attractions here.

Choosing and fitting a radio/cassette deck

Like Clarion, Kenwood have gone to great lengths to create a 'family resemblance' across their product range. The KRC 452L is a well-featured, mid-range deck, with 2 x 25W (4 x 15W) max output, quick-out, Dolby NR and CD changer control.

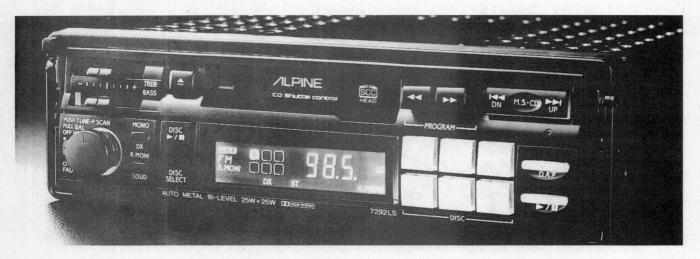

Masters of deck identification are Alpine Electronics; those large, green, easy-to-use buttons are an instant giveaway. Their cassette decks and tuners alike have been praised from all quarters and the have an ability to equip sets with multi-function controls which are still simple to operate. With 2 x 25W output, quick-out, auto metal, the 7292LS can also operate Alpine's 6-disc shuttle CD autochanger.

Pioneer has a wide span of sets, typified by the KEH 4200, with its rounded, safety conscious look and large, easy to use, controls. Another quick-out set, it features Dolby 'B' NR, Best Station Memory and 2 x 25W power output.

Choosing and fitting a radio/cassette deck

Nakamichi have a reputation of manufacturing the best cassette decks in the world. To some, they produce a better sound than Compact Disc. Whether that is right, only your ears can tell you. The TD 700C is typical of the simple, straightforward design adopted by the company. Whilst quality is always guaranteed, so is a high price label which ensures that these decks are found only in the cars of the true enthusiasts.

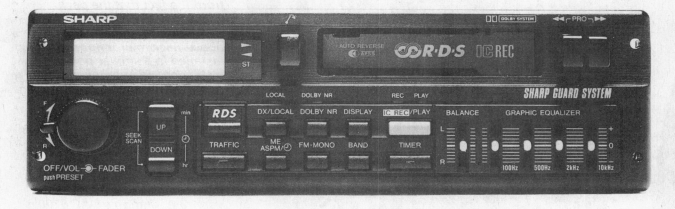

Almost all manufacturers now produce at least one RDS set, which means that this excellent facility is available throughout the price spectrum. This Sharp deck reflects their penchant for in-built graphics and a generally high level of features.

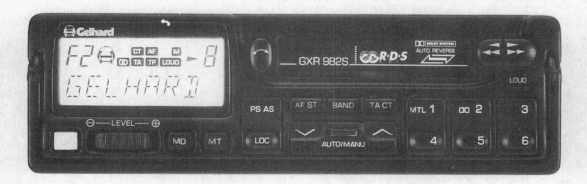

Gelhard are a fairly new name to the UK market, but RDS sets feature in the range. Note the alpha-numerical display on this GXR 982S, one of the best features of RDS - you can see what radio station you're listening to!

Choosing and fitting a radio/cassette deck

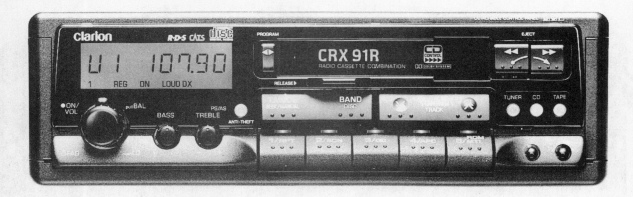

The Clarion CRX 91R is a well specified RDS deck with twin security features and 4-channel line-out, making it ideal for use on its own or with separate amplifiers. This set is shown in more detail in the 'fitting' section of this chapter.

Sanyo claimed a world first with the WX22, initially unusual for being a double-DIN sized set. However, its originality comes from the fact that the deck accepts instructions by spoken word; the installation of a goose-neck microphone, usually installed in a similar position to a standard cellular telephone microphone, allows the driver to pre-programme his personal command words. As such, he can literally talk to the set which will then do his bidding.

Yes, DAT players still exist, but they are very expensive and very rare. Clarion's DAC 8100 is typical of the breed looking, with its digital counters etc., like a cross between a CD player and a standard tape deck which is, in essence, what it is.

Choosing and fitting a radio/cassette deck

RDS

RDS is a must for anyone who listens to the radio a lot and/or travels across the country. Early RDS sets (like CD players) were massively expensive, but prices are falling rapidly and it won't be long before most sets are so equipped.

RDS is an abbreviation of Radio Data System and takes the form of an inaudible digital code. When this is 'piggy-backed' onto the normal FM signal, it is capable of providing extra services with suitably equipped sets.

The BBC commenced their five-year plan of RDS development in 1988. Claimed to be the greatest step forward since the transistor, RDS has a lot to live up to. The features listed here are not available on all RDS sets, the specification of which are likely to vary enormously with cost and with age. When buying, you should check the specification very carefully before parting with your cash.

CLOCK DATE/TIME

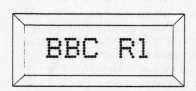

A suitable receiver will display a very accurate time and date. Moreover, it will automatically take account of variations such as British Summer Time or world time zones.

NETWORK INDICATION

Network Indication one of the best RDS features. The set displays the name of the station being received rather than just the frequency, eg BBC R1, as shown here. It is particularly useful for drivers who travel around the country, where retuning can leave them without a clue as to which station they are listening to.

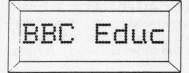

It will also show when you are listening to a programme which is part of a network split. For example, where Radio 4 is broadcasting educational programmes on FM but normal programmes on the Long Wave, the display would read as shown here.

AUTOMATIC SAME NETWORK FOLLOWING SYSTEM

RDS will search out and transmit the best frequency for a particular station in any given area. Thus, as you drive around the country, the programme reception will remain at a consistently high level, without the need for constant retuning. This diagram shows how the system would work for a vehicle travelling between Norwich and Peterborough. When leaving Norwich, the Tacolneston transmitter would be the one being received. By the time the car reached King's Lynn, the signal would be deteriorating and the tuner would be 'looking' for something better. In this case, it would find that the Peterborough transmitter offered a far better signal and would switch to it automatically. Such changes are made in a manner inaudible to the listener. As can be seen, the Tacolneston transmitter covers East Anglia whilst Cambridgeshire, South Lincs and parts of Leicestershire are all covered by the one at Peterborough.

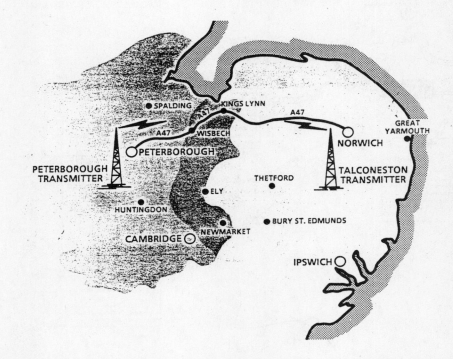

Choosing and fitting a radio/cassette deck

FULLY AUTOMATIC TUNING
All you need to do is select the name of the station required and the tuner will find it for you.

TRAFFIC INFORMATION RECEPTION
You don't have to travel far in the UK (particularly on our congested motorway systems) before you meet some roadworks. When you've been held up for an hour or two, you may start to think about getting an RDS equipped tuner! The traffic information service is one where the current sound source (tape/CD etc) is interrupted and traffic news broadcast instead. When the particular flash has finished, the machine returns you to your original choice. These travel bulletins are localised, even if you are actually listening to a national station. This diagram shows a typical application of the RDS travel service and how it can help the driver to avoid potential 'snarl-ups'.

EON
EON is a development of RDS and stands for Enhanced Other Network. Its main object is to allow traffic information broadcasts from local radio stations to be available to you, even if you are listening on a national radio network.

FUTURE DEVELOPMENTS
The services available now are the tip of the RDS iceberg. Possible future features include; programme type selection where, at the touch of a button, the tuner would seek out the type of programme, rather than any specific programme) you wanted to listen to. Possible examples are pop music, educational, news etc. The system would be able to handle up to 30 different types of programme.

PROGRAMME ITEM NUMBER
Missing the start of a radio broadcast is easy to do if you are concentrating on driving (I hope you are!). With PIN, you would be able to program the tuner to switch on to (or change to) a specific station at a specific time, much like a video recorder.

```
CBSO Conducted by David Atherton
Vaughan Williams-London Symphony
```

Radio text. If you've ever tuned into a broadcast half way through and not known who or what you were listening to then radio text is for you. The set's display panel would show exactly what was being broadcast at the time. A 'message' of up to 64 characters in length could be displayed, meaning that it could be used also for correspondence addresses etc. However, its use for in-car audio is expected to be limited, because of the obvious safety aspect.

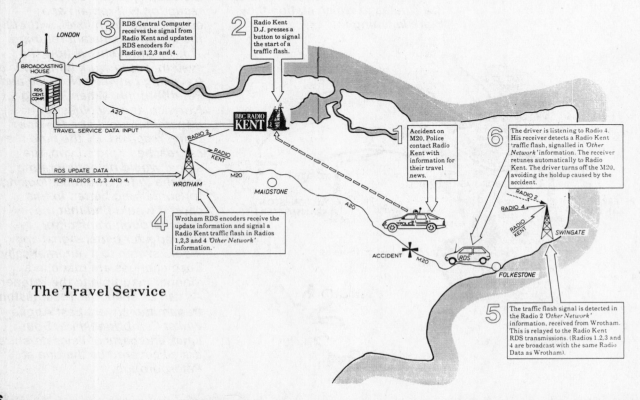

The Travel Service

Choosing and fitting a radio/cassette deck

MUSIC/SPEECH BALANCE
The sound balance of transmitted programmes from the BBC is a compromise, bearing in mind the wide variety of sets and environments in which it is received. With this feature, the in-car listener will be able to adjust the volume of speech and music separately, in order to achieve a precise balance. .

BROADCASTING FREQUENCIES BBC RADIO

Station name	Medium Wave Metres	Kilohertz (kHz)	VHF(FM) Megahertz (mHz)
Radio 1	275/285	1053/1089	88/90.2
Radio 2	330/433	693/909	88/90.2
Radio 3	247	1215	90.2/92.4
Radio 4	1515(LW)	198(LW)	92.4/94.6
Bedfordshire	258	1161	96.9
Bristol	194/227	1548/1323	95.5
Bournemouth	221	1359	-
Cambridgeshire	207/292	1449/1026	96.0
Cleveland	194	1548	96.6/95.8
Cornwall	476/457	630/657	96.4/97.3/95.2
Cumbria	397/206/358 7	56/1458/837	95.6/96.1
Derby	269	1116	96.5/94.2
Devon	351/303/206/375	885/990/1458/801	97.5/97/96.2/103.9
Furness	358	837	96.1
Guernsey	269	1116	-
Humberside	202	1485	96.9
Jersey	292	1026	88.8
Kent	290/388	1035/774	96.7
Lancashire	351/193	855/1577	96.4/103.3
Leeds	388	774	92.4/95.3
Leicester	358	837	95.1
Lincolnshire	219	1368	94.9
London	206	1458	94.9
Manchester	206	1458	95.1
Manx Radio	219	1368	96.9/89.0
Merseyside	202	1458	95.8
Newcastle	206	1458	95.4/96.3
Norfolk	351/344	855/873	95.1/96.7
Northampton	271	1107	96.6/103.3
Nottingham	197/189	1521/1584	95.4
Oxford	202	1485	95.2
Radio Cymru			92.4/94.6/96.8
Radio Scotland	370/513	810/585	92.4/94.6/97.6
Radio Wales	457/340	657/882	-
Reigate	219	1368	104.0
Sheffield	290	1035	97.4/88.6
Shropshire	397	756	96.0
Ludlow	189	1584	95.0
Solent	300/221	999/1359	96.1
Stoke on Trent	200	1503	94.6
Sussex	202/258	1485/1161	95.3/103.1
West Midlands	206/362	1458/828	95.6
York	450/238	666/1260	90.2/97.2

Choosing and fitting a radio/cassette deck

INDEPENDENT LOCAL RADIO

Station name	Medium Wave Metres	VHF(FM) Kilohertz (kHz)	Megahertz (mHz)
Birmingham B/C	261	1152	94.8
Beacon B/C	303	990	97.2
Capital	194	1548	95.8
Cardiff	221	1359	96.0
Chiltern; North	378	792	95.5
South	362	828	97.6
Classic FM	–	–	100.3/101.9
County Sound	203	1476	96.6
Devonair; Exeter	450	666	95.8
Torbay	314	954	95.1
Downtown Radio	293	1026	96.0
Essex Radio;			
North	220	1359	96.4
South	210	1431	95.3
Gwent B/C	230	1305	104.0
Hereward;			
Peterborough	225	1332	95.7
Northants	193	1557	102.8
Invicta Sound	242	1242	103.8
London B/C	261	1152	97.3
Manx Radio	219	1368	86/96.9
Marcher Sound	238	1260	95.4
Merica Sound	220	1359	95.9
Metro Radio	261	1152	97.0
Moray Firth	271	1107	95.9
North of Scotland	290	1035	96.9
Piccadilly Radio	261	1152	97.0
Plymouth Sound	261	1152	96.0
Radio Aire	362	828	94.6
Radio Broadland	260	1152	97.6
Radio City	194	1548	96.7
Radio Clyde	261	1152	95.1
Radio Forth	194	1548	96.8
Radio Hallam	194	1548	95.2/95.9
Radio Luxembourg	208	1440	–
Radio Mercury	197	1521	103.6
Radio Orwell	257	1170	97.1
Radio Tay;			
Dundee	258	1161	95.8
Perth	189	1584	96.4
Radio Tees	257	1179	95.0
Radio Trent	301	999	96.2
Radio Victory	257	1170	95.0
Radio West	238	1260	96.3
Radio Wyvern;			
Hereford	314	954	95.8
Worcs	196	1530	96.2
Red Rose Radio	301	999	97.3
Saxon Radio	240	1151	96.3
Severn Sound	388	774	95.0
Signal Radio		1170	104.3
Southern Sound	227	1323	103.4
Swansea Sound	257	1170	95.1

Choosing and fitting a radio/cassette deck

Station name	Medium Wave Metres	VHF(FM) Kilohertz (kHz)	Megahertz (mHz)
Thames Valley B/C	210	1431	97.0
Two counties Radio	362	828	97.2
Viking Radio	258	1161	102.7
Virgin Radio	1215/1224/ 1197/1242	–	–
West sound	290	1035	96.2
Wiltshire Radio; Swindon	258	1161	96.4
W Wilts	321	936	97.4
West Yorks Radio	362	828	94.6

enlarge it. Great care and thought is required here. The first point to bear in mind is that the value of the car (especially if it is a 'classic') may be reduced if you take a hacksaw to it. Also, you must ensure that nothing else (wiring, control cables etc.) is damaged as you enlarge the hole.

If you have no DIN aperture in the dash (or if it is occupied by another piece of ICE equipment) then life gets trickier. A centre console is the next best position, either one supplied by the manufacturer or an aftermarket version. However, you must secure the console firmly to the car and the deck firmly in the console for it to work at its best. Last and definitely least, you could 'sling' the deck beneath the dashboard. Here, it is least handy for driver operation and is more difficult to hold securely in place.

MOUNTING POSITIONS

Without a doubt, the best place for your radio/cassette deck is in the standard DIN aperture in the dashboard of your car. If you're lucky, the manufacturer will have sited it carefully, ideally within easy reach of your left arm and as close to the top of the dash as possible, so that you don't have to take your attention from the road for too long. If you have an ICE aperture which does not suit your new set, then the odds are that it is ISO DIN and in order to be of use, you will have to

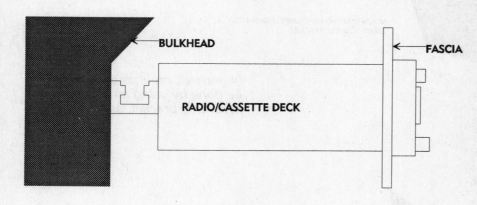

If you have such an aperture in the dash, you must make sure that the set is supported at the rear. Using a bracket like this is one way, though many cars actually have a bolt already in the bulkhead for just this task. An alternative—

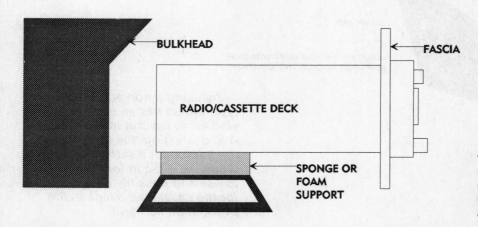

—is to support the deck underneath, by using a sponge or foam support. Resist the temptation to let the set rest on a pile of cluttered wiring.

Choosing and fitting a radio/cassette deck

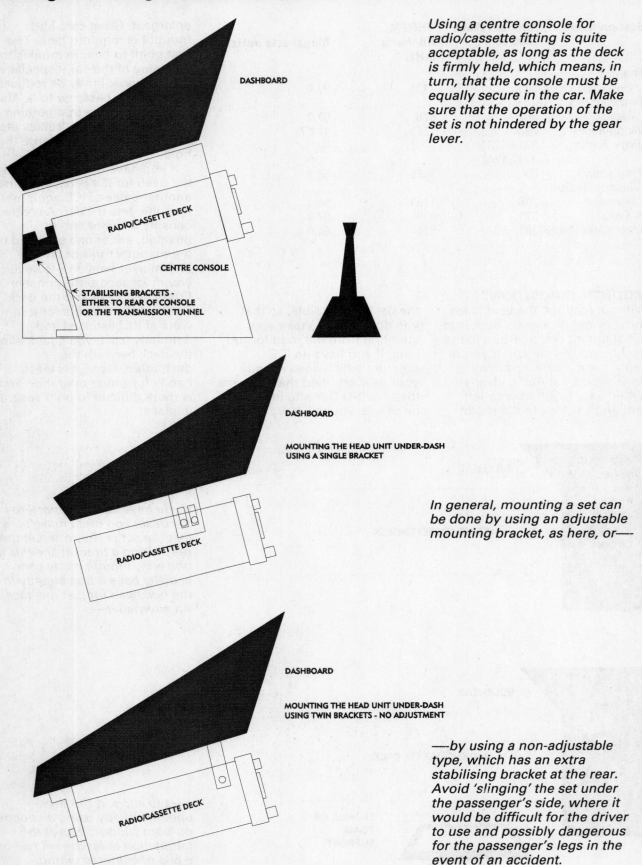

Using a centre console for radio/cassette fitting is quite acceptable, as long as the deck is firmly held, which means, in turn, that the console must be equally secure in the car. Make sure that the operation of the set is not hindered by the gear lever.

In general, mounting a set can be done by using an adjustable mounting bracket, as here, or—

—by using a non-adjustable type, which has an extra stabilising bracket at the rear. Avoid 'slinging' the set under the passenger's side, where it would be difficult for the driver to use and possibly dangerous for the passenger's legs in the event of an accident.

Choosing and fitting a radio/cassette deck

The 'Before you Start' Checklist

1. Make sure that your set is complete with everything the manufacturers say it should have and that you will not need to purchase anything specifically to fit it.
2. Ensure that the set will physically fit where you want it to go. DIN size sets fit DIN apertures, but anywhere else will need some ingenuity and possibly a mounting bracket or two.
3. Read the instructions thoroughly and don't start until you fully understand them.
4. Have you got all the terminals you need? You may need to swap some (either set or vehicle).
5. Have you got some spare fuses, just in case?
6. You'll need an aerial. Refer to chapter 4 and consider that, if your aerial is any age, now is perhaps the time to uprate to match your brand new, hi-tech, hi-fi.
7. Make sure that your speakers are up to their new job, especially if you've increased your power output.
8. Make sure that you have enough time to complete the job with some in hand for a spot of fault-finding.
9. Have you got enough room to work? Fitting equipment in cramped conditions is irksome, to say the least. In some cases, it makes life much easier to remove the steering wheel to gain access to the rear of the DIN aperture.

In this section, a Clarion CRX 91R is seen being fitted at Dards Electronics. One great advantage of going to a specialist (or even buying your set there and 'picking his brains') is that you can glean a few tips. For example, this particular Ford Escort has a built-in 'joystick' fader on the dash, which makes 4-speaker wiring a little more tricky than usual. Wiring the deck to the fader as the makers intended is possible, as long as you are not running a high-power system. If you are, or if you just want to use the deck fader, then you will have to locate the rear speaker cables at the lower kick panels by the driver/front passenger's feet and the front speaker cables under the dash. Adapting the wiring from the original fader isn't possible.

The set shown here is particularly well qualified and would serve you well, sounding good on its integral amplification with the promise of even better things to come if extra amplification were linked up to the 4 line-outs and a 6-disc CD autochanger plugged into the back of the deck. Its features include RDS tuning, Dolby 'B' noise reduction, 4 x 15W (max) output, key-off pinchroller release and LCD display. RDS facilities include; TP (Traffic Programme Identification) PS (Programme Service Name); PI (Programme Identification); TA Traffic announcement and AF (Alternative Frequency). Such a deck would prove more than a little temptation to a thief and so there are two security features in evidence; CATS III (Computer Anti-Theft System) security coding and a removable front panel, which drops nicely into its own pouch and from there easily into a jacket pocket or handbag. The latter is becoming increasingly important, as there is a marked tendency for those with quick-out decks to leave them in situ because of the bulk of the set. A nice touch with the code system is the option to make the anti-theft indicator a flashing LED, the better to warn off a potential thief **before** he breaks your window. The option to uprate is important on any set if you enthuse about your in-car sounds.

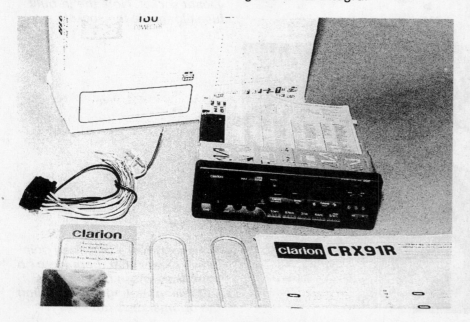

Even the professionals start by laying out the set, the ancillary bits and pieces and the instructions to make sure that everything is present and correct. Reading the instructions before you start is vital, it gives you an overview of the job in hand and what you will need.

Choosing and fitting a radio/cassette deck

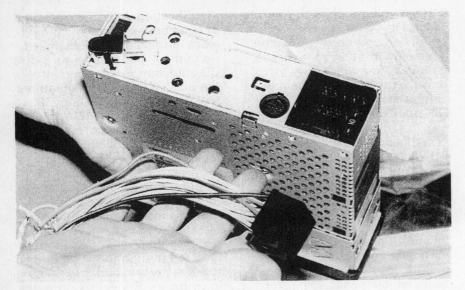

Unlike a full quick-out deck (where the wiring is taken to a cage fitted into the dash), the wiring in this case is taken directly to the set. However, life is made easier by the fitting of a multi-pin plug, which means you don't have to complete complex wiring with the set hanging out of the dash on your knee.

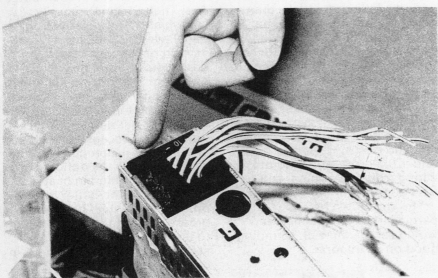

The plug provided is designed for use when you are powering up your speakers from the deck's integral amp. If you want to use separate power amplifiers, you will need a special adaptor to fit in the top, vacant socket. Note the in-built fuse alongside the socket.

The Escort has a DIN sized hole which means that there were no complications in finding a suitable mounting place. Pulling the original deck out—

Choosing and fitting a radio/cassette deck

—revealed someone else's wiring. You should be prepared for the worst; most of the time you'll be right and if you're not, it'll be a nice surprise! Although it looks confused, this particular set was by no means as bad as some.

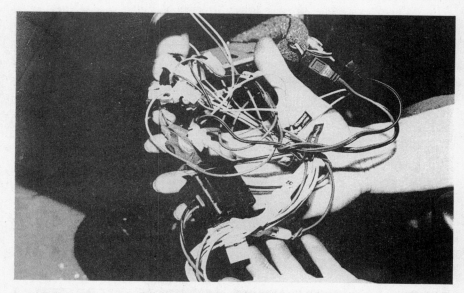

The Ford fader was equipped with DIN speaker plugs. The set already in the car was fitted with suitable sockets, but often there will be a difference at this point. This leads to a choice - to alter the set to the car or the car to the set? Unless your car is outrageously valuable (in which case, doctor the deck) then there's not a lot to choose. But do use suitable connectors, suitably fitted.

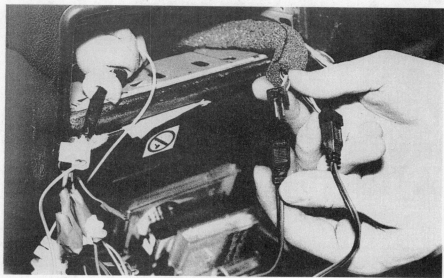

Whatever the state of the wiring, resist your initial impulse to rip it all out wholesale. You first need to check it through to see what connects what. It's a good idea to use masking tape to label the various wires (front speaker, accessory power etc) if they haven't already been identified in some way. The last thing you need to do before you remove the battery is to check out which wires are carrying power, with and without the ignition on. As a rule, the deck should be powered through an ignition fed live, with a permanent live being used to feed the memories. This reduces the possibility of leaving the deck on overnight and flattening your battery. The S/P digital tester is the modern equivalent of the electrical screwdriver but is more convenient for most in-car tasks, having no wires to get in the way.

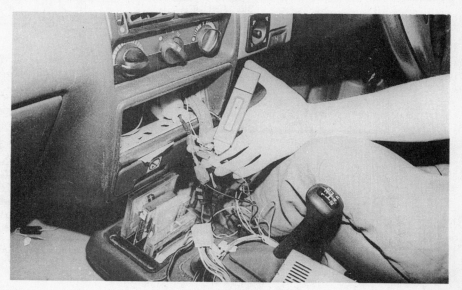

Choosing and fitting a radio/cassette deck

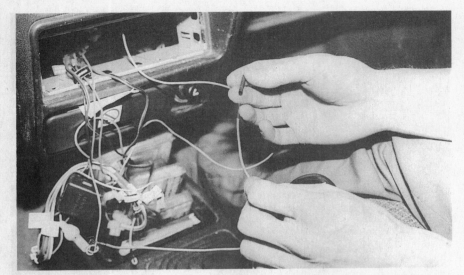

As with most professionals, Barry preferred to solder electrical joins rather than use a Scotchlok or other connector. If you can, follow his lead; if you can't remember that a good crimped or Scotchloked join is better than a bad soldered one. Before soldering, he slid a heat shrink insulator over one of the wires (it won't fit on afterwards!) and...

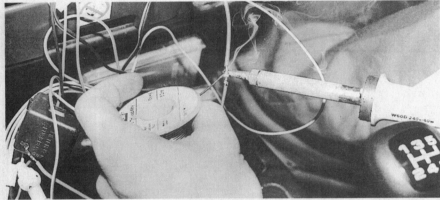

... twisted the bared ends of the wires in question together before applying the iron to the join. Once it was sufficiently hot, the applied solder ran easily through the join.

The other wiring was completed in an equally professional manner, guided by the clear schematic diagram supplied with the deck. (Courtesy Clarion Shoji (UK) Ltd)

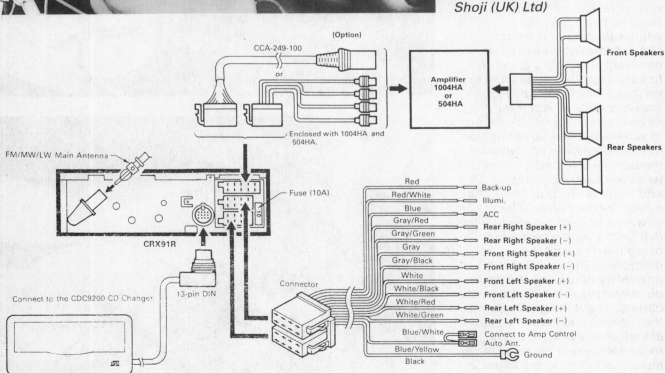

Choosing and fitting a radio/cassette deck

The deck is secured in the DIN aperture by means of a metal bracket which slots into position and...

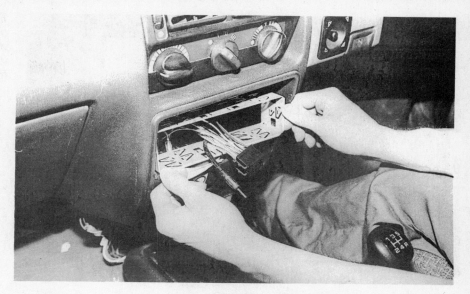

Using a strong screwdriver is the usual method, but make sure it can't slip and damage your person.

All that remains then is to slide the set neatly into the aperture, but before you do so, it's a good idea to switch on and make sure everything is working as it should. This is often where you find that you have forgotten to plug in the aerial lead!

Choosing and fitting a radio/cassette deck

The Clarion deck is aesthetically pleasing, with a design which will sit at ease in most cars. It won't please the thief, though for...

... the detachable front panel is a simple yet effective method of immobilising the deck. Once unclipped, it leaves a bright, vacant panel and the words 'Security Code' awaiting a potential illegal owner. With the four-figure code enabled as well, it's a set not worth stealing.

The only way for the owner to remove the deck once it is in place is to use a couple of special 'keys' which fit into holes on either side of the deck and release the locking mechanism.

Chapter 6
Choosing and Fitting an Aerial

In order to get the best from your radio tuner, you'll need a good aerial. As radio waves whiz about the air at an incredible 300,000,000 metres per second, you'll understand that trying to capture them and feed them to your deck is no small task.

The aerial (or antenna) has to convert the electro-magnetic waves into a signal which can be understood by the tuner which in turn has to convert them into ones which can be passed onto the speakers. Not only that, but it has to do this while the car is moving about and the distance to the transmitting station is constantly varying. All this while the British weather attacks it with extremes of snow, rain wind and, occasionally, sun!

In theory, any aerial should be one quarter of the length of the radio wave being received. Therefore, if the set were tuned to Radio 2, then the aerial would need to be one quarter of 330 metres. Elementary maths shows us that the aerial would need to be 82.5 metres long! Obviously, this is far from being a practical proposition and even if it were, you'd have to change the aerial length whenever you wanted to change stations. So, thanks to advanced circuitry in most modern tuners, an aerial with a length of around 100 cm is quite adequate.

The basic choice is of telescopic aerial, an electric version of the same thing, a roof/'A' pillar mounted unit or a windscreen model.

The telescopic aerial (in either format) is the most popular, with the usual fitting position being the front wing. Though a roof mounted aerial will, pound for pound, give a better signal, they are less popular, not least because the fitment of a sunroof can cause problems and because fitting is more difficult because of the need to route wiring beneath the headlining.

Windscreen aerials generally compensate for their lack of 'height' by incorporating a signal amplifier. Most are stuck across the top of the windscreen and have the obvious benefits that they cannot be vandalised (not unless the whole screen is broken!) or left out to be savaged by the car wash.

The 'before you start' check list
1. Make sure that the aerial you are to fit is complete. Compare what you have in the box with the manufacturer's list. If something is missing, you need to find out **before** you start work.
2. Ensure that the aerial is suitable for your car before you buy it. Check he amount of room beneath the wing, especially if you are fitting an electric version where the motor will require more space than usual.
3. If you are fitting a roof aerial, check the distance between the roof and headlining and that the routeing of the cable will present no insurmountable problems. The same principle

Choosing and fitting an aerial

The Hirschman aerial is designed as a roof aerial but can also be mounted on an 'A' pillar. In practical terms, most roof aerials are not retractable, which leaves them open to vandalism or the car wash. However, this model can be screwed off the mounting bracket and placed inside the car for safe keeping.

applies to 'A' pillar mounted units.

4. Read the instructions and make sure that you understand them. Ensure that you can identify the various items. If you are fitting an electric aerial, can it be linked into your present head unit to become totally automatic (i.e., the aerial extends when the unit is switched on and visa-versa) or will you need to fit a separate switch?

5. Allow enough fitting time, including some for a spot of fault finding.

6. Tools for the job. An average aerial fitment will require;
A drill, pilot bit, varicut bit.
Torch,
Medium slotted/crosshead screwdrivers.
Pliers
Hexagonal spanner (usually 17mm)
Insulating and masking tape.
Crimper and/or soldering iron.
Anti-rusting agent.

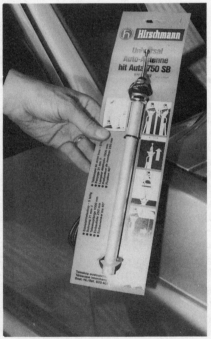

The wing-mounted telescopic aerial is the most popular choice and many cars come so-equipped direct from the vehicle manufacturer. This model is the Hirschman universal and is lockable. The mounting place would usually be in the front wing, although mounting it in the rear wing (of a front-engined car) would place it further away from the interference-laden engine bay. However, it would mean that the cable would have to be extended to reach the deck and the routeing would create a lot more work, with plenty of opportunities for mistakes (trapping the cable in the rear seat, for example). If you do have to lengthen an aerial cable, I would suggest that you use a proprietary extension lead, with pre-wired plugs.

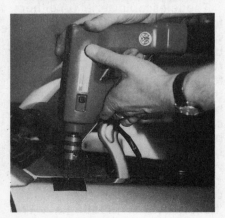

The instructions give clear details of the hole which must be drilled in the wing to accommodate the aerial. Use masking tape on the wing so that the hole position can be marked with a biro or pencil and so that any drill slippage will not damage the paintwork. Make the first hole using a small pilot bit and follow this up with progressively larger bits or—

Choosing and fitting an aerial

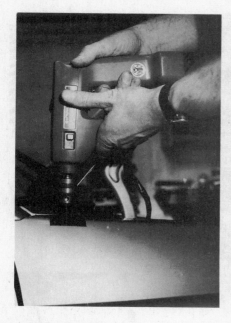

—- much simpler, with a single, Sykes-Pickavant varicut drill bit.

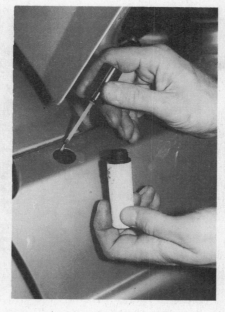

The next step is very important, apply some anti-rust agent to the bared metal edges, otherwise the wing will start to rot. There needs to be some metal contact between the aerial and the underside of the wing. In order to achieve this but prevent rusting, use something in the order of Duckhams Copper Ten, a copper based grease which allows electrical conductivity whilst keeping out moisture.

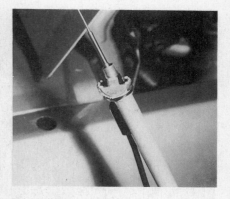

The aerial has to be part-dismantled before fitting into the wing. The lower spherical half stays on the stem, which has to be inserted from the underneath. This aerial includes a metal stabilising bracket, which is used to prevent the lower section of the aerial straining the rest. In most metal-bodied cars, this would not be necessary, though on cars with fibreglass bodies, it would probably be advisable.

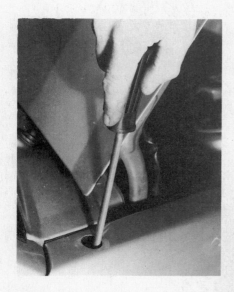

Whenever you drill a hole in metal, you'll end up with burrs (rough edges) around the new hold. Prevent possible injury by using a round file to remove them.

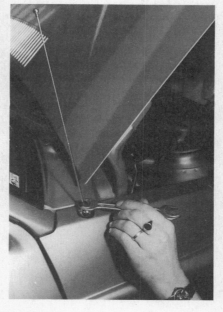

When it is in position, the sealing washers and hexagonal nut can be added and tightened up. The aerial lead should be passed through into the car, preferably avoiding the engine bay altogether. Many cars have a hole in the body pressing specifically for an aerial lead. If you have excess cable, tie it neatly together (but avoid kinks). Do not cut aerial cable. Make sure that the cable cannot interfere with the workings of such items as windscreen wiper mechanism.

Choosing and fitting an aerial

When the aerial is pushed all the way down, it can only be extracted by using the two-pronged key, making it a good security measure.

FITTING AN ELECTRIC AERIAL

The increasing popularity of the electric aerial is not hard to understand. Initially, there was a certain element of one-upmanship, as the aerial slid serenely up and down of its own accord (there still is to a certain extent) but as prices have come down, its use has become more widespread and the practicalities become apparent. Most of us have, at one time or another, joined a busy motorway, switched on the radio and only then realised that the aerial is down. Similarly, in the carwash is not the best of times to remember that the aerial is still extended. Some aerials require a separate switch so that the driver chooses when to extend or retract. However, many modern sets have a built-in lead which, when connected to a suitable electric aerial, facilitates fully automatic operation; when the deck is switched on, it extends, when the deck is switched off, it retracts. As well as the benefits already mentioned, it is also an excellent security measure; it means you can't leave your aerial up (and a target for vandals), even accidentally.

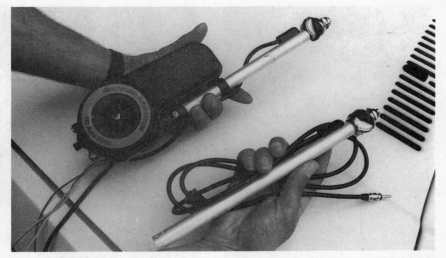

The difference between the manual and electric units is clear in this photo. The electric requires much more depth and width than its manual cousin.

As can be seen, it requires much more room beneath the wing to cope with the large motor. Also, it needs to have a stabilising bracket. It is wise to check before you purchase it that it is suitable for your particular car. Not all universal aerials are truly universal.

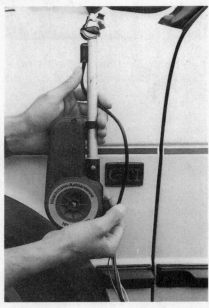

The Hirschman aerial comes complete with all necessary wiring which has terminals ready connected where necessary in its own wiring loom. A relay is also included.

Choosing and fitting an aerial

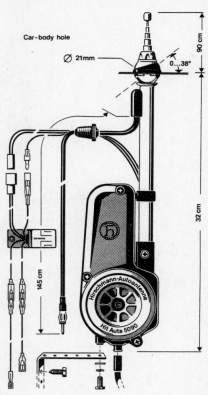

The comprehensive instructions include a layout diagram showing how fitted should achieved and the various dimensions. (Courtesy Hirschman Ltd.)

Also included is a simple-to-understand schematic wiring diagram. As with the manual unit, a grommet is fitted to the aerial lead for fitment where it comes through the inner wing. This prevents dirt and moisture coming through into the car and guards against the possibility of the lead being wrenched and damaging the aerial. (Courtesy Hirschman Ltd.)

Key to diagram
1 - 6   Antenna base components
7   Drain tube
8   Screw
9   Screw/spring washer/lock washer
10   Stabilising bracket
11   Plastic grommet
12   Relay
13   Connector

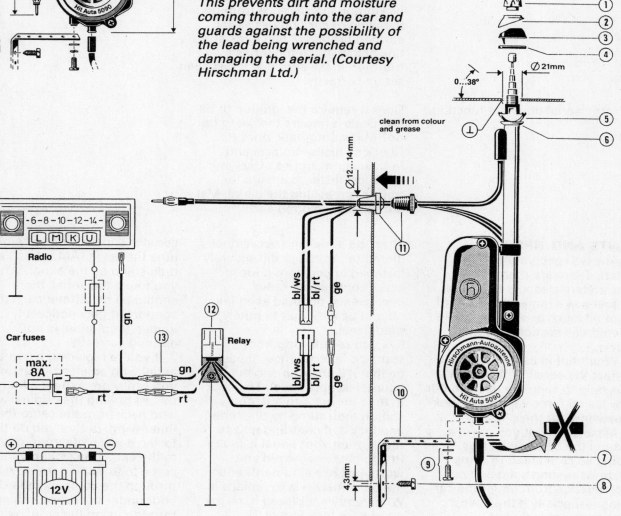

101

Choosing and fitting an aerial

FAULT FINDING

Symptom	Remedy
Set switched on, but no sound except static hiss	Check set in 'radio' mode (combination units). Ensure aerial lead is connected to set. Ensure aerial is fully extended.
Crackling noises, whines, hums, etc.	Probably car borne interference - see chapter 7. Ensure aerial fully extended. Faulty telescopic sections. See hints and tips below.
Poor reception	Listening area is important, especially with FM. Tall buildings etc. can make a big difference to the signal. Ensure the set is correctly tuned to a strong signal and that the aerial is plugged in firmly and not just making intermittent contact. Ensure aerial is fully extended, especially the lowest section which is prone to picking-up interference from under the wing or car body. Is aerial earthed properly? See hints and tips below. If the tuner has a trimmer, make sure that it has been set up correctly.
Electric aerial does not function	Does it require the ignition to be on? Does it require the set to be on? Many automatic ones do. Check all fuses - vehicle and in-line, aerial and set. Use your electrical tester to ensure that power is reaching the aerial. Make sure there is a good earth.

HINTS AND TIPS

1. Always keep your aerial clean. There are many proprietary cleaners available, or just use a rag and a squirt of light oil (such as WD40) to help keep those sections moving freely.
2. When not in use, you should retract the aerial. Not only will this help to keep the dirt out, but it will also force vandals to look elsewhere for their 'fun'.
3. Make sure that you always extend the aerial fully. Not doing so could lead to some sections rusting solid and/or interference from under the car wing, especially if the lowest section is not included.
4. Check soon after you have fitted the aerial and occasionally thereafter, that it is still securely fastened in position. A loose aerial could cause radio interference or could even fall off and be a danger to other road users.
5. If you're suffering from sporadic 'interference', it could be that it's not your set, but your telescopic aerial. To check it, tune the set into a strong station then stand by the aerial and flick it. If crackling noises result, then your aerial is faulty (the joints become old and worn, just like a human!) and the only answer is to replace it. When you've replaced it, take a good hard look at hint 1!
6. A quick way to find out if your aerial is earthed correctly is to tune the set in (AM is best) and then stand by the aerial. When you touch the aerial, there should be no difference in the sound. If it gets noticeably louder, your aerial is not earthed correctly.
7. If you're replacing a manual aerial with an electric version you can score a great own goal by just pulling the original aerial lead back from the set to the inner wing. Before you do this, fasten a piece of string or wire to the end. Thus, when you come to threading the new lead through the complex innards of your under-dash, you can pull it through using the string/wire.

Chapter 7
Dealing with Interference

The in-car radio tuner has progressed in leaps and bounds over the past thirty years and shows no sign of slowing down. Along with such obvious improvements as electronic tuners, liquid crystal displays, auto-seek, RDS etc., the manufacturers have also been working hard on other, less obvious areas. One of these is interference suppression, where Philips, for example, have excelled with their much-copied IAC (interference absorption circuit) systems.

However, in-car tuners are up against it in a big way and you should not be too surprised if you get some form of interference, especially if your set or car (or both) are fairly old. Before attempting any form of interference suppression, it is important that you have a good quality aerial, correctly fitted and earthed. Most forms of radio interference come from various electrical items already fitted to the car. In some cases (plug leads, for example) they could be faulty. In others, they may be working perfectly, but just need suppressing. Listed here are some of the most common sources of interference and the types of noise they are likely to make.

With any car, the metal bonnet provides an excellent screen for unwanted noise. However, to be totally effective, it must be well earthed. If your bonnet does not have an earthing strap, then you should fit one. Most accessory shops will be able to supply a universal item to suit. If it does have one, then you can lose nothing by removing it, cleaning

Item	Noise
Clock (analogue)	Regular ticking, even with ignition and engine switched off.
Windscreen wipers	Crackling sound when in use.
Heater fan motor	Crackling, as above, or a whine.
Windscreen/rear screen washer motor	Whine
Fuel pump	A fast irregular ticking, fast and then slow.
Coil/plug leads/plugs	A crackling sound which rises and falls with engine revs.
Alternator (or dynamo)	A whine in unison with the engine revs.

103

Dealing with interference

up the contact areas with emery paper and smearing a little Copper Ten on both contact surfaces before replacement. In successfully suppressing any source of interference, the first task is to isolate the culprit. Essentially, it is a task which requires patience and a logical mind to work steadily through the various stages of isolation until the offending item(s) is found. To carry out the following checks, it is important that the car be parked in a reasonably open space, otherwise confusing signals could get in the way (next door's CB radio etc). Also, the doors, hatch/bootlid and bonnet should be shut.

Start by sitting in the car with the engine and ignition off and the radio on. Most in-car electrical items require the ignition on for operation and so interference at this point will almost certainly be the clock. Switch on the ignition and, one by one, switch on all the electrical items (wiper motor, heater fan) you can think of and listen for interference. If interference occurs simply by switching on the ignition, it points to a fuel pump problem. If you cannot locate the problem, switch on the engine and listen for the sounds given in the table listed earlier. All modern cars are fitted with spark plug caps with internal suppressors. Over a period of time, these fail and cause interference. Another common cause of complaint is the distributor, which often develops hairline cracks in the cap or a loose or worn rotor arm. Straight replacement is the answer. Similarly, fitting a metal shield around it can sometimes help; indeed, some vehicle manufacturers now fit these as standard equipment.

It could also be some non-standard electrical item that you have fitted as an 'extra' to your car. Listen for this in the same manner as above.

If you are getting interference from only one speaker or pair of speakers, it could be that the wiring loom is causing it via the speaker cables. Re-routeing the speaker leads usually eradicates the problem. Ensure that the speaker cables do not run close to any power carrying leads or electric motors. It is also possible for the radio's power lead to induce interference. This is usually solved by selecting a different power source.

Often, an interference problem becomes evident after fitting an uprated system. However, it does not mean that the problem is new - it could have been present for years, but has only become noticeable given the extra capability of the new system.

Manufacturers are working hard to equip their sets with lots of in-built resistance to interference, however caused. These micro-chips are part of Clarion's latest generation of anti-interference circuitry.

You should note that FM (VHF) signals are much more prone to interference than AM signals. Take extra care when earthing the set and a roof aerial will help to get a better signal. Because of the nature of FM signals, they tend to bounce off hard objects and in certain causes, you could find that your tuner is receiving the same signal twice. A reflected signal could take much longer than the original to reach your set and, as such, there could be distortion and fading. There is not a lot you can do about this, although, because the car is constantly moving, there is more than a probability that the car will soon move away from whatever is causing the reflection.

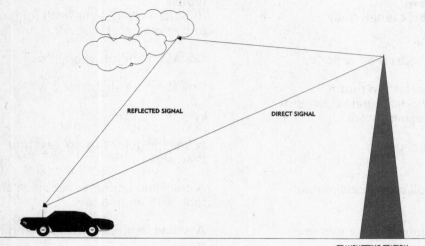

HOW FM SIGNALS CAN BE REFLECTED AND CAUSE INTERFERENCE

Dealing with interference

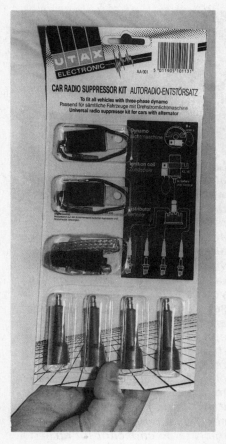

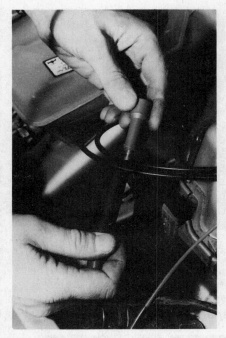

Fitting the spark plug suppressors is easy. Pull of the spark plug cap, push fit the suppressor into it and—

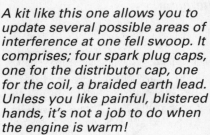

A kit like this one allows you to update several possible areas of interference at one fell swoop. It comprises; four spark plug caps, one for the distributor cap, one for the coil, a braided earth lead. Unless you like painful, blistered hands, it's not a job to do when the engine is warm!

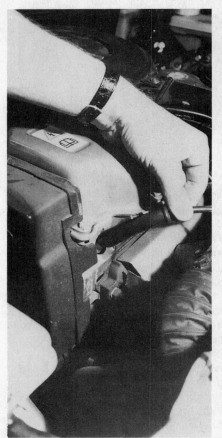

—-replace the whole lot onto the plug. Don't mix the plug leads up otherwise your car will be reluctant to run. Always replace one lead at a time. The caps have a tendency to be a tight fit on the leads and so a healthy tug may be necessary to get them off. Don't forget to give a similarly healthy push when you replace them. Remember that the leads themselves are also subject to wear and tear and that after a few years, it could be that their suppression qualities are on the wane. Replacing a set of plug leads with a new set of silicon replacements may well be a good idea if your car is any age.

If you have fitted the suppressors and replaced the leads but are still getting interference which sounds as if it is coming from the HT circuit, it could be the plugs themselves. Some plugs have in-built suppressors and can be identified as they have the letter 'R' as a prefix or suffix. Again, replacement of plugs is a simple task and it could well solve your problems.

Dealing with interference

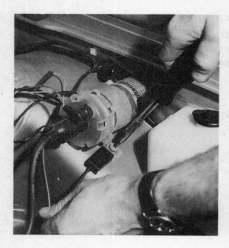

Fitting a suppressor to the coil is equally simple. The suppressor has to be securely earthed by its mounting bracket. The best way is to place it under the coil mounting brackets. The size of capacitor depends on whether your radio has FM waveband (most do, nowadays). For FM sets, the capacitor rating needs to be 2.5 microfarad but for AM sets it need only be 1 microfarad.

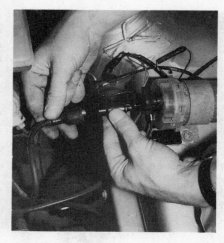

The high tension lead to the coil can also be covered by using the in-line suppressor, which is fitted in the same way as the spark plugs.

Later model cars tend to have alternators whilst earlier models were equipped with dynamos. Both are a common cause of radio interference. Some alternators will have capacitors already fitted, which should be visible and situated close to the other electrical connections. All you can check is that it is securely held and that the leads are firmly attached. If all appears to be well, but you still think that it is the root of the problem, then any good motor electrical dealer should be able to test it for you quite cheaply. If it is found to be faulty, the it should be replaced. use a 3 microfarad capacitor for an alternator.

The vehicle bonnet makes an excellent screen (as long as it's a metal car) but it must be earthed to the car body. This kit includes a strong metal braid for this purpose. In this case it was not required, as the vehicle was so fitted as standard. However, many cars are not equipped with one and it could make a big difference to your radio reception, especially on FM. Whether you are fitting a new one or checking the old one, make sure that there is a good contact at both ends by cleaning the metal surfaces with emery cloth and adding a dab of Copper Ten to prevent the build-up of corrosion.

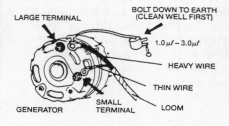

The suppressor lead then has to be connected to the ignition side of the coil.

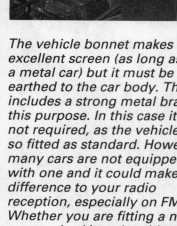

This diagram shows the correct way to install a suppressor on a typical dynamo. As usual, it is vital to ensure that there is a very good earth connection. If you are dealing with the type of dynamo where it is not possible to use the original terminals, then you will have to Scotchlok into the nearest lead. Loosen the nearest mounting bolt and slide the capacitor under it. Once more, a task for the fine emery cloth, for cleaning both the unit itself and the capacitor. use a 1 - 3 microfarad capacitor.

Dealing with interference

SUPPRESSION OF MOTORS

Having established which motor is causing the problem, the first task is to ensure that it is earthed correctly. Where it is earthed through its casing, remove the motor and clean up its earthing point and that on the bodywork with a fine emery cloth. If this has no effect, try running a separate earth wire from the motor to a known good earth. If the problem persists, you will have to fit a either a 1 microfarad capacitor or a 7amp, in-line choke or, in extreme cases, both.

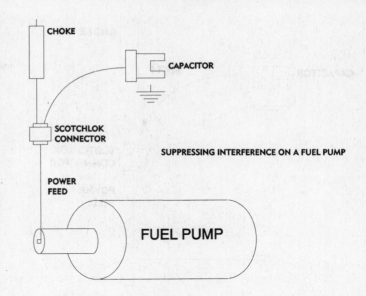

A fuel pump should be suppressed in the manner shown here, with a capacitor and/or a choke, and a clock—

The capacitor lead needs to be connected into the power lead of the motor. In some cases, it is possible to use a 'piggy back' spade connector. In others you will have to use a Scotchlok connector. The capacitor should be tightened securely to a nearby earthing point, as shown in this diagram.

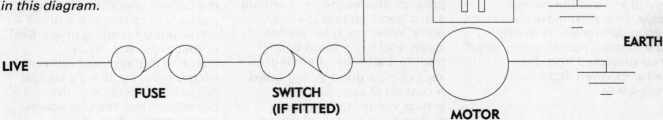

An in-line choke is basically a small coil of wire which produces a magnetic field when fitted in the supply lead of a component. This effectively cancels out interference. It should be fitted as close as possible to the item in question. It is fitted in a similar manner to a capacitor, usually by Scotchloking into the power lead and then attaching a good earth.

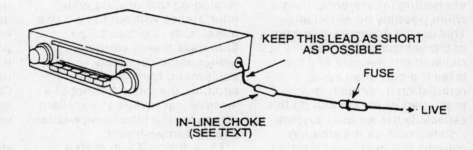

Dealing with interference

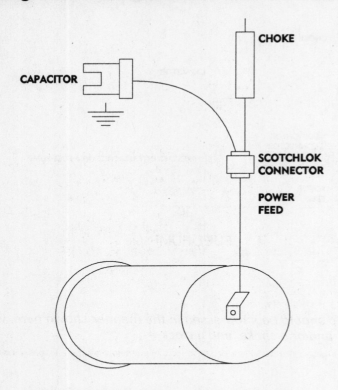

SUPPRESSING INTERFERENCE ON AN ELECTRIC CLOCK

—can be treated in a similar way. Don't forget that whilst using Scotchloks is a simple way to achieve the desired result, you must make sure that a good connection is made. Bear in mind also that they need to be protected from the elements when fitted in the engine bay.

If you have suppressed everything but still find that you are getting interference, then it could possibly be re-radiation. This occurs when another part of the vehicle picks up the radiated interference and the takes the part of an aerial, reradiating it. Virtually any metal part of the car can do this, especially the exhaust system.

Items such as the steering column and gear lever are also potential problem areas. Find out which is the culprit by going around the vehicle with an earthing strap. Try each possible source in turn, earthing it to a good, known chassis point. When the interference stops, you have found your trouble. Earth the offending item using a good strong braid. A coating of Copper Ten or similar will protect against corrosion and retain electrical conductivity.

FIBREGLASS CARS

Owners of cars with fibreglass bodywork (Lotus, TVR, many kit cars etc.) will not need reminding that dealing with interference without the aid of a metal body is difficult. The fibreglass makes earthing (either aerials or audio equipment) hard work. In addition, the metal bonnet of a 'normal' car makes an excellent screen for the interference-laden engine compartment.

Thus, fitting ICE in such a vehicle demands more of the fitter than usual. You must make sure that the equipment you fit has a good earth. The best way is to take a separate braided lead directly to the chassis. Mounting the aerial away from the engine bay is much more important than in a metal bodied car. If possible, mount it at the rear of a front engined car and the front of a rear engined car, again, ensuring a really good earth. If you have to extend the aerial lead, I would recommend using a proprietary extension with pre-fitted plugs/sockets. For power, take it direct from the battery, remembering, of course, to maintain the presence of an in-line fuse.

In the engine bay, some manufacturers fit a small metal shield over the coil etc. If you do not have one, it is worth contacting the makers to see if it is available as an official spare. If not, you could look to making one up yourself. However, where and how to mount it is likely to be a problem. Clearly, it must not foul on the engine or the bonnet when closed. It is possible to regain some of the interference repelling properties of that metal bonnet by adapting a fibreglass bonnet accordingly. By fitting a special foil to the underside of the bonnet, some screening against interference will be gained.

It may be possible to glue it in place or possibly fibreglass resin could be used. You must make sure that there is absolutely no possibility of it falling off whilst the car is in motion, as this could be very dangerous. Many ICE specialists keep a supply of specially perforated metal for specifically for this purpose. The foil you use to cook your turkey at Christmas is not suitable, it will cook your engine in much the same manner!

Having fitted the foil, you should ensure that it has a good earth to the chassis.

Chapter 8
Compact discs – Description, Care and Storage

Following their not inconsiderable success with the Compact Cassette, Philips developed the Compact Disc, or CD. The digital sound produced by a Compact Disc when played on a good machine (there are a few that are not much better than a good cassette deck) is far superior to that of a standard compact cassette tape, in terms of quality and dynamic range.

As long as the disc is treated with some respect and the playing surface not scratched, it will last almost indefinitely. A five year old cassette, played on a regular basis, would be showing more than a few signs of wear and tear. A five year old disc would still be playing perfectly, certainly mine are! The iridescent compact disc itself is around 12 cm in diameter (there are smaller 'single' discs) and can usually store around 75 minutes of music. Unlike a conventional record, it produces its sound from one side only, (the underside, without the writing on). The whole disc is covered in a layer of protective plastic, thicker on the underside for obvious reasons.

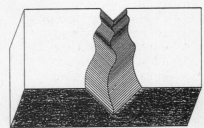
L.P. RECORD GROOVE

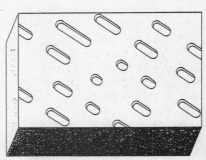
CD SIGNAL "PITS"

The compact disc begs comparison with an LP record, where a stylus is used to trace mechanically along an analogue groove in the vinyl disc. This is subject to all manner of outside influences such as dust, finger prints and scratches. (Diagram courtesy Alpine Electronics UK Ltd)

There is no mechanical contact at all with a compact disc, which means that, if cared for, the disc will always give the same, high quality sound. Unlike the LP record, a beam of laser light is used to 'read' digitally encoded information from the disc. The information is stored in billions of microscopic 'pits' on the surface of the disc. It is standard procedure on the better CD players to use 3 beams; one which actually plays the disc whilst the other two run either side of it and detect even the slightest deviation from the correct path. Any such errors are corrected in milliseconds and are usually undetectable by the listener.

Compact discs, description, care and storage

CARE OF YOUR COMPACT DISCS

Handle your CD collection as little as possible. When not in use, keep them in their original cases, special carriers (such as Fischer C-Box or Caselogic) or in the magazine of a CD autochanger. When you do have to handle them, keep your fingers away from the playing surface, in much the same way as you would with an LP record. Place the forefinger in the centre hole and steady the disc with your thumb.

Yamaha designed this single disc cartridge system which has been used by Blaupunkt and Clarion. With the disc installed in the cartridge, it can be handled much as you would a disc in its own plastic case. However—

This is how not to hold your discs! As well as the possibility of scratching the disc this way, the dirt and grease from your fingers can confuse the lasers and cause inaccurate reading of the disc.

—when it comes to loading, there is no need to remove the disc. The cartridge is loaded as-is, into the machine, where a special mechanism lifts a flap at the front and removes the disc for playing. When the cartridge is ejected, it reverses the procedure. Though a good idea, it does make the discs tricky to store in the car and you have to have a specific type of machine.

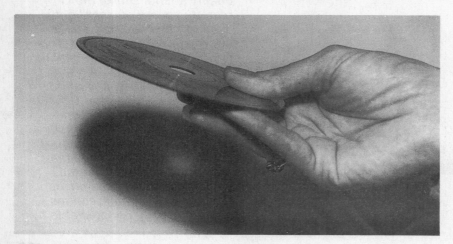

Compact discs, description, care and storage

It's important for your discs to be clean at all times, for grease (even that small amount from your fingers) could contaminate the playing surface and lead to skipping and jumping of certain tracks. The Allsop disc cleaner is...

... a simple, yet rather clever, idea for keeping your discs in tip-top condition. The lower half of the machine is covered with foam to protect the disc, whilst the smaller upper half features a revolving, foam pad which in turn is linked via a system of cogs...

... to a handle on top. With the disc sandwiched as shown here, and the cleaning solution (supplied with the kit) applied, the turning motion of the handle cleans the disc like new.

IN-CAR DISC STORAGE

Compact discs are inherently more difficult to store in-car than cassettes; they are physically larger and much more delicate. Whilst in their boxes, they can be handled with relative ease, but they will take up a lot of space. Also, it means that when you change discs, you have an awful lot of messing about to do. Trying to get a disc out of its box whilst making sure the disc coming out of the machine doesn't get damaged and then loading that one into its own box demands four hands and should not be attempted whilst driving. If you

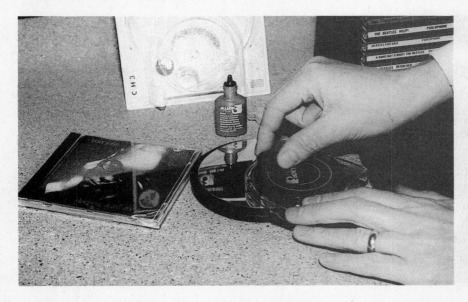

have an autochanger, you have less of a problem, for at any given time, you are effectively 'storing' six or more discs within the machine itself. Extra discs (should you be journeying trans-Europe!) can be carried in any of the carriers shown here. If you have an autochanger of the 'magazine' type (e.g. Pioneer, Clarion etc.) then would be well advised to invest in a second cassette to carry around your second six discs.

111

Compact discs, description, care and storage

There are few natural storage places for compact discs in a car; you can usually get one or two in the map pockets (if your car has them) and sometimes, the shape of the glovebox will allow some to go in there. But far better than all these is to go for a purpose made CD carrier. There's quite a lot of choice and Fischer C-Box have, as you might imagine, come up with a stylish range of CD holders to match their long-established range of cassette boxes. This 4-disc version is DIN size and fits into any standard aperture. In this case it is fitted into an after-market centre console. As a direct alternative—

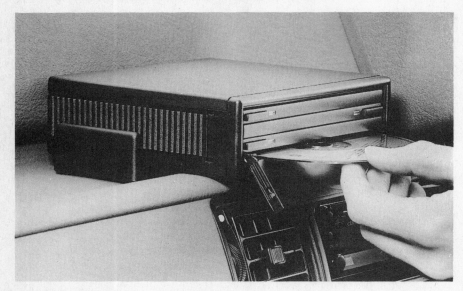

—this version relies on the same mechanism, but is supplied with a bracket for mounting wherever is convenient, either vertically or horizontally.

Such is the increasingly popularity of the in-car compact disc player, that Fischer are producing a range of custom-made CD holders in the same style as their renowned cassette boxes. This six-disc version fits exactly in the centre console of the MKIII Golf.

Compact discs, description, care and storage

The alternative to 'permanent' in-car storage is a transportable method of carrying the discs. The Caselogic carrier shown here meets very definite criteria to ensure that your precious (and expensive) discs stay in one, cared for, piece. This particular model holds 15 discs (in their original cases) with the facility for one of them to be a 'double'.

The strong, zippered case keeps everything together when not in use.

113

Chapter 9
Choosing and fitting a CD/Tuner and Autochanger

As you will read in chapter 8, the compact disc offers inherently better sound quality and greater longevity than a standard compact cassette. As well as this, there are other advantages to CD. The average cassette holds several hundred metres of tape and it can often take several minutes to wind from one end to the other. This same procedure takes three or four seconds on CD. Swapping backward and/or forward to find favourite tracks is just as quick, a definite bonus when the driver's attention needs to be on the task of driving. As with the cassette, it was introduced first for domestic use and its popularity there soon ensured that a healthy in-car market sprung-up.

Many of the first in-car CD players were fractious items, with a tendency to be highly strung and disc skipping being the order of the day. In addition, it was some while before technology advanced to such a stage where a tuner could be added and it took even longer to be able to fit an amplifier into that DIN size machine. Nowadays, the Compact Disc player has very much come into its own, with highly specified machines being available at a fraction of their somewhat archaic (now) looking forebears. At the time of writing, you can buy a well featured radio/cassette deck, with integral 2 x 25W amplifier and a 6-disc CD autochanger for 2/3 of the price of a CD player only (i.e. no radio, tape or amplifier) in 1987; now that's what I call progress!

SPEAKERS
An in-car CD player will demand more of your speakers than a cassette or tuner and you will almost certainly need to upgrade in order to get the best from your digital sound source. In extreme cases, it is possible for the wide dynamic range to damage speakers which are not up to the job. If you have not got component speakers fitted (a mid-woofer and tweeter combination, for example) then you should consider doing so now. Remember that CD players have a very wide frequency response and not having speakers able to match it will, to some extent, negate the fitting of a CD player in the first place.

AMPLIFICATION
Most CD players (or CD/tuners) come with some form of integral amplification, usually around 50W. Ideally, you should look to have in the region of 100W for any CD-based, 4-speaker system in order to make the best of the CD player's dynamic capabilities.

WOW AND FLUTTER
Even the very best cassette decks have some element of wow and flutter - a slight variance in the tape speed. It has to be particularly bad for it to be noticeable, but on a CD

Choosing and fitting a CD tuner and autochanger

player, wow and flutter is so small as to be below measurable limits.

WHAT TO LOOK FOR
Tuner features
Overall, tuner facilities offered on CD/tuners reflect what is available on a similar radio/cassette decks - refer to chapter 5 for further details.

CD FEATURES
Because of the nature of CD players, there is little required other than the standard controls - items such as Autoreverse (so useful on a cassette player) are clearly inappropriate.

OVERSAMPLING
Oversampling relates to the way in which the digital sound is produced. Basically, the more times oversampling you have, the better the sound. The first CD players were 2 times oversampling, though this moved quickly on to 4 times, 8 times and 16 times. In truth, whilst some difference may be detectable in the quiet of your living room on a domestic player, you'd be hard pressed to detect the difference between 8 and 4 times oversampling once you're on the road.
Nevertheless, the more times oversampling the machine, the better the quality sound.

The FFWD/FREV buttons move up or down a track and on many machines, holding these down allows you to run through a track at double speed in order to find a particular section. Flipping from track to track in seconds is one of the great joys of using a CD player.

The REPEAT function allows you to reply either a specific track or the whole disc.

DISPLAY is usually either the time left on the disc or the elapsed time of a specific track (with the option to toggle between the two).

RANDOM PLAY (or MIX or SHUFFLE or variants of the two) is a feature whereby the machine will play the tracks in no specific order. On an autochanger, this means that you could start with track 6, disc 4 and follow it with track 12 disc 2 etc. Its main use is where a listener has few discs to choose from and gets fed up of hearing the same old order over and over again. This function is also available on many single disc players, though its operation is more of a novelty value. The better RANDOM/MIX functions will not repeat a track and, therefore, are not truly random, but, in this case, you really don't want them to be.

SCAN has the same function as when it is applied to either tape or tuner. When pressed, the machine will play 10 secs of the first track and then move onto to 10 secs of the next one. This will continue until you find a track you want to listen to.

MEMORY is a feature available on some decks. It allows the user to preset the order that tracks are played in, either just on one disc or, in the case of an autochanger, on a number of discs. Because of the 'planning' element involved here, many owners do not use the function.

Last, but by no means least, you should look for some form of security; many CD players are now available in 'quick-out' format or with detachable front panels. Whatever, a good CD/tuner is not cheap, even though it may be good value and buying a deck for someone else to listen to doesn't make much sense.

WHICH TO CHOOSE
Most manufacturers produce a CD player of some description, with many offering a wide range of variants. The three basic CD player formats, one of which is the multi-changer which follows later in this chapter:

CD PLAYER ONLY
As the name suggests, just a CD player, usually in a DIN size box. Some have an integral amp, though many rely on plugging into the amp(s) already in your system. They are not the most popular of machines, for most people would prefer a combination unit, in much the same way that cassette players are usually sold with radios. If you use the DIN aperture in the dash, it means that you're going to be pushed for space if you want to add a tuner and/or cassette deck as well. The reverse is true; if you're adding a CD player to your existing set-up, whereby the DIN dash aperture is already occupied by a radio/cassette deck, you will probably end up mounting the CD player under the dash.

CD/TUNER
The CD/tuner is the digital equivalent of the radio/cassette deck and far more practical than a CD player on its own. For many, it has been well worth while to swap from a radio/cassette deck to CD/tuner, though this tends to depend largely on the number of CDs owned, a factor linked to domestic CD player ownership.

The main disadvantage the CD/tuner shares with the CD player is that loading single discs is a tricky manoeuvre and, though it should not be attempted whilst driving, many people do. Some CD/tuners can be linked to a multi-disc autochanger.

As with radio/cassette decks, you should also consider the appearance of a set as well as its technical merits. After all, you will be spending a lot of time looking at it, and if it is totally unsuitable for your car's interior, it's going to take some of the pleasure away from CD listening.

Choosing and fitting a CD tuner and autochanger

The Sanyo FXD 601 has a 2 x 25W (4 x 15W) output, line-out, 4 x oversampling, 24 preset memories and quick-release security.

Like the Sanyo set, Kenwood have gone for a standard, radio/cassette type appearance for their KDC 74D. It features 8 times oversampling, 32 presets, quick-out security and 2 x 15W power output.

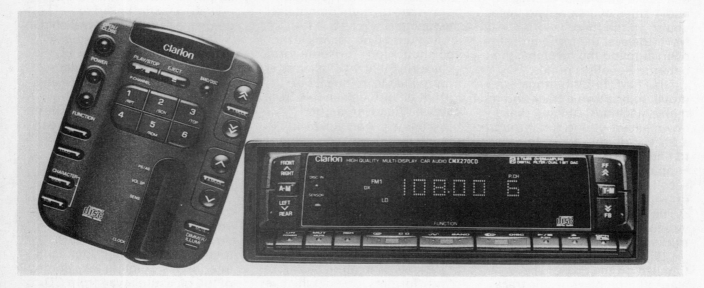

The Clarion CMX 270CD leaves no one in any doubt that it means business, with 8 times oversampling, infra red remote control and CD autochanger controller capability. The unusual front panel covering the CD aperture slides open and can be set at whatever angle is required for easy reading of the display.

Choosing and fitting a CD tuner and autochanger

The Alpine 7906R is a top-of-the-range deck, featuring an RDS tuner, 8 times oversampling and a whole host of other features, along with security quick-out facility. The general look of the deck is mirrored in the rest of the company's products, such as...

... the 7980, an interesting hybrid deck, which combines a 3-disc multi-disc changer in a DIN size head unit.

117

Choosing and fitting a CD tuner and autochanger

An alternative to mounting a CD player permanently in your car is to opt for a portable CD player, such as the Clarion CDC1005. This is a handy unit which can be used wherever you are and can be powered by a mains transformer—

—dry cell batteries, which slot into this pop-out section at the rear or by means of a special adaptor from a vehicle cigar lighter.

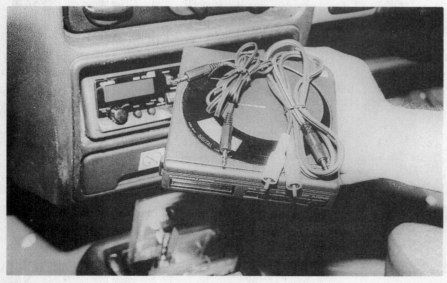

This model comes with 2 line-out feeds, which means that it can be plugged into a head unit fitted with a suitable socket or separate amplifier. Naturally, many Clarion units have suitable sockets, though they are by no means alone.

Companies such as Goodmans now produce purpose-made devices to secure such players in-car. The important points to remember are that it should be safe (i.e., not likely to fly around the car in the event of sudden braking or an accident) and it must be securely held, otherwise you are likely to have problems with skipping tracks etc.

Choosing and fitting a CD tuner and autochanger

Carrying your Portable CD player is best achieved by using a purpose-made case, like this one from Caselogic. The strong padded case holds the player and there is a pocket capable of carrying a couple of CDs and/or the various leads and attachments. An adjustable carry strap is also supplied. Note that a zippered hole in the bottom of the case means that the machine can still be used whilst in the case.

SAFETY WARNING: Please do not use personal stereo headphones whilst driving, it isn't safe!

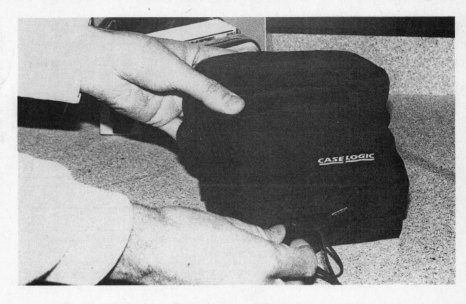

The 'Before you start checklist'
1. Make sure that your set is complete with everything the manufacturers say it should have and that you will not need to purchase anything specifically to fit it.
2. Ensure that the set will physically fit where you want it to go. DIN size sets fit DIN apertures, but anywhere else will need some ingenuity and possibly a mounting bracket or two.
3. Read the instructions thoroughly and don't start until you fully understand them.
4. Have you got all the connections you need? You may need to swap some (either set or vehicle).
5. Have you got some spare fuses, just in case?
6. If you're fitting a CD/tuner, you'll need an aerial. Refer to chapter 4 and consider that, if your aerial is any age, now is perhaps the time to uprate to match your brand new, hi-tech, hi-fi.
7. Make sure that your speakers are up to their new job - CD in particular makes heavy demands on them.
8. Make sure that you have enough time to complete the job with some in hand for a spot of fault-finding.
9. Have you got enough room to work? Fitting equipment in cramped conditions is irksome, to say the least. In some cases, it makes life much easier to remove the steering wheel to gain access to the rear of the DIN aperture.

TOOLS FOR THE JOB
Fitting an average CD player or CD/tuner will require;
Screwdrivers (slotted/crosshead)
A crimper and/or soldering iron
12v test lamp or multimeter
Torch
Pliers
Insulating tape
Electrician's/masking tape
If you are fitting a specific bracket, you will need and electric drill and relevant bits.

If you are fitting an autochanger, you may well find a trim removal tool useful, especially if you are mounting it at the rear of the car.

This 'fitting' section was carried out at Dards Electronics, Milton Keynes, where I watched how it **should** be done.

The Philips set as it comes. It's well featured (as you will see from the specification at the end of this chapter) and ergonomically sound; that is, the controls are arranged logically, easily operated and look good. Note that a small wiring loom, connectors fuses and everything else required for fitting are included with the machine.

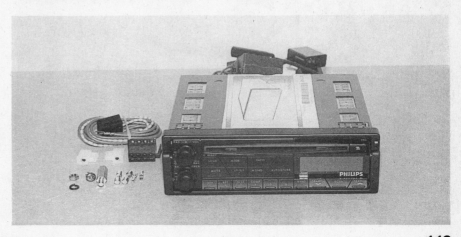

119

Choosing and fitting a CD tuner and autochanger

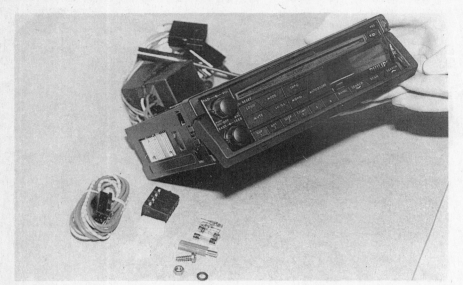

The DC980R is a removable set for security purposes. The handle, seen here, allows it to be carried easily (though a Caselogic case would make life simpler still) and when the deck is inserted back into the cage, it locks securely and safely into place.

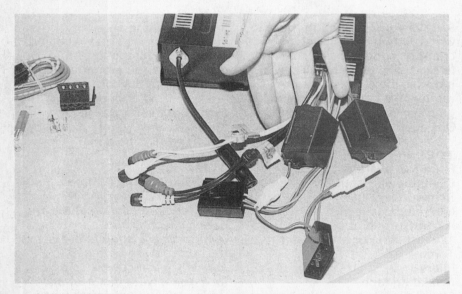

All connections are made to the back of the cage, rather than the set itself. The wiring here is not particularly complex, though you will clearly need quite a lot of space behind the deck to get it all in. Those two large 'boxes' are full of anti-interference electronics.

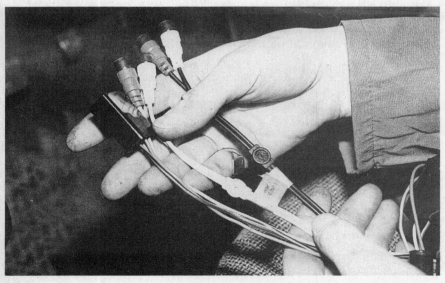

Philips make wiring easy and logical. The four RCA connectors are the line-out sockets and are used only if you are adding extra power amplifiers. Note that they are fitted with plastic caps; if you are using the set only with its standard, 50W integral amp, then leave the caps in place and tie the wiring neatly out of the way. The options are either to rely totally on the integral amp, or use it to power the front with a power amp for the rear or ignore it altogether and use power amps for both front and rear. The other socket accepts a plug which, when wired, takes the power, earth, accessory and speaker cables.

Choosing and fitting a CD tuner and autochanger

As with most ICE equipment, there is a comprehensive instruction manual (in several languages) complete with easy to grasp, schematic diagrams. This, then is the first task - read it!

Once you know what is going to go where, remove the earth terminal from the battery, the only caveat being those cars with clever computers and/or engine management systems. Check in your vehicle handbook or your supplier before you go any further.

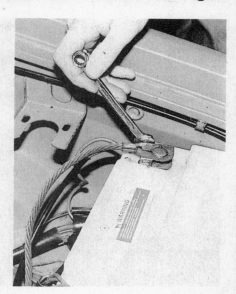

If the deck already in the DIN aperture is a factory-fitted unit, then you will find lots of neat wiring and correctly wired plugs. If, however, someone else has been DIY-ing before you, it's anyone's guess what sight is going to greet you. This radio/cassette deck had apparently been wired by someone with a serious interest in spaghetti! Untidiness is forgivable (just) but—

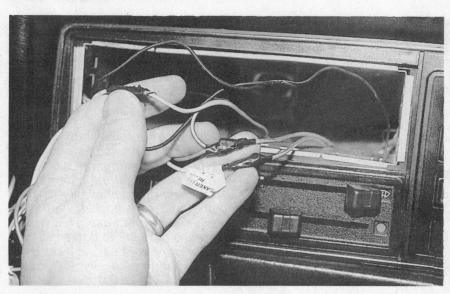

—-using electrician's tape like this, is not. As you can see here, wires aplenty were twisted together and covered with tape. This is not only bad practice, it is potentially quite dangerous, as you could easily end up with a power carrying lead wafting around under the dash, short-circuiting everywhere. Don't forget, you use electrician's tape to insulate a join and to loom cables together, not to make a join.

Choosing and fitting a CD tuner and autochanger

If the deck you are removing has a metal housing fitted, you will have to take this out, too. Prise back the securing lugs which fix it in the dash.

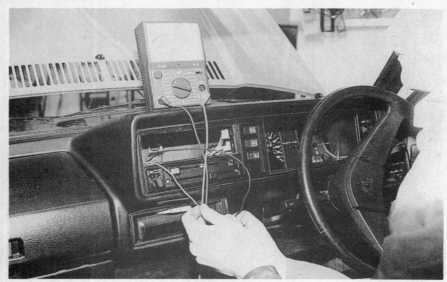

Now is the time to temporarily replace the battery earth terminal in order to check which wires are which, especially when faced with wiring such as we've seen earlier! Dard's professional fitter used a multimeter, though, for the most part, a test light is sufficient.

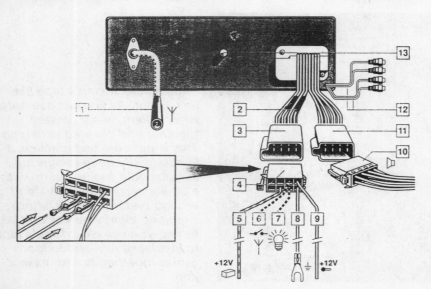

This schematic diagram shows how the set should be wired. Note the 'illumination' wire - a most useful feature, whereby whenever the vehicle lights are on, the deck is backlit, even when it is not in use. This means that, in the darkened interior of your car, you can see to switch in on easily. (Courtesy Philips Car Stereo)

Choosing and fitting a CD tuner and autochanger

Now this is the way to make a join! The butane gas powered soldering iron, makes in-car soldered joins easy and—

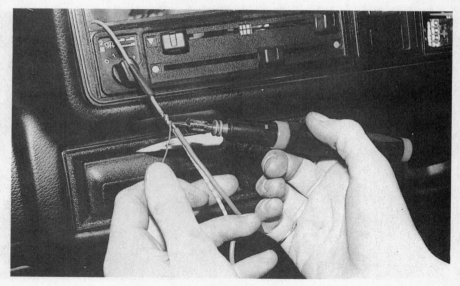

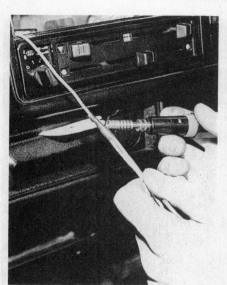

—by adding a suitable attachment, the heat shrink cover can be 'warmed' into place. Again, the heat shrink cover is a professional preference and whilst they cost only pence and are very useful, electrician's tape will do much the same job.

All of the wiring goes into this multi-plug and—

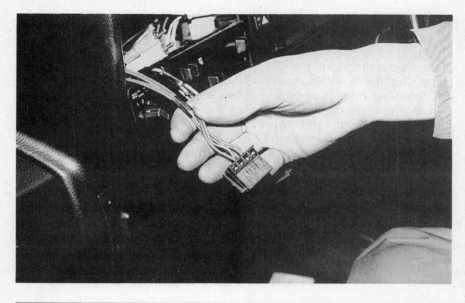

—the wires are held by simple, screw connectors. Make sure that you use the correct sized screwdriver, otherwise you could chew the heads up.

123

Choosing and fitting a CD tuner and autochanger

The DIN cage slots into the aperture and is secured by *bending back* the lugs. Make sure that the wiring is neat and tidy and not trapped in any way.

The finished item. The subject of ergonomics can be raised again here. The almost totally black finish of the Philips deck blends well into the interior of this car.

**Philips DC980R
Brief specification**
Digital tuner, FM/MW/LW
3-beam CD player
CD insertion with automatic start
Dynamic range compressor (CD)
Loud
4-channel line-out
Quick-out security chassis
2 x 25W (max) power output
36 presets with Autostore
DX (local)
4-channel line-out
2 x 25W integral amplifier

THE MULTI-DISC AUTOCHANGER

Originally seen as just a way to spend a lot of money on an unnecessary item, the cost of autochangers has plummeted since its introduction. Added to their current value for money is their practicality. As long as you have somewhere to mount one safely and securely, they are incredibly useful for the avid music fan. By loading a magazine of (usually) 6 or 10 discs, you can have up to twelve hours of music literally at your finger tips, and changing discs no longer involves an undignified (and possibly dangerous) fight with the CD case. Just press a button and the 'changer does the rest.

Autochangers can be operated either on their own or via a dedicated radio/cassette deck or CD/tuner; in the case of dedicated head units, the cassette/CD & tuner controls 'double up' to work the CD player.

If you go the 'dedicated' route, then it becomes easier on the pocket, as the head unit can be installed and operated on its own whilst cash is accumulated to add the autochanger. Usually, there is very little in the way of extra wiring required to fit an autochanger. Whether you go for a 10/12 disc autochanger or one of the increasingly popular, 6-disc versions, is largely a matter of available space. The latter can often be fitted in the glovebox or under the seat, making magazine swapping particularly easy.

THE FM MODULATOR

If you have a radio/cassette deck at the head of your system, you may want to add a CD player/changer but still retain the original head unit. If the

Choosing and fitting a CD tuner and autochanger

deck is dedicated to an autochanger, there is no problem, just plug it in. There was little in the way of options until the arrival of the FM modulator. Effectively, the CD player (either single or autochanger) is fitted and wired as usual but instead of being connected to an amplifier/speakers, it is plugged into the FM modulator, which in turn is linked to your aerial.

Also plugged into the FM modulator is the radio/cassette deck. So, when the deck is tuned to a suitable frequency (in tuner mode) whatever is playing on the CD will come through the system. If you have a video recorder linked to your TV, you will understand the principle. The only drawbacks are that the sound quality is dependent on the quality of existing equipment being used (aerial and tuner) and on the quality of signal being received.

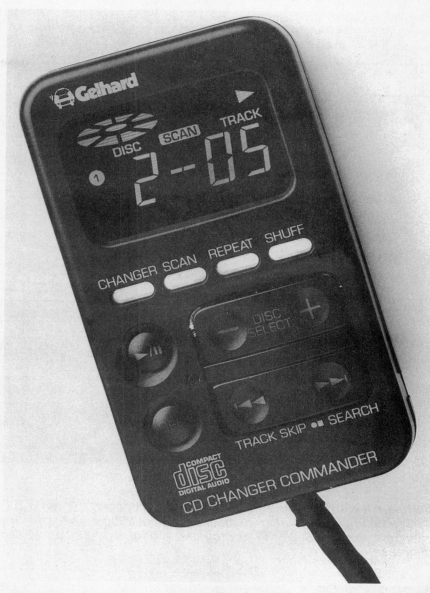

Most autochangers can be operated as separate units, with their own controls. The Kenwood method, with the KCA R20M is to use two units, one for the function controls with the other being a display module.

Gelhard prefer to put everything into one box. Usually, there is the option to mount such controls permanently or, alternatively, keep them in the glove box for use when required.

Choosing and fitting a CD tuner and autochanger

Kenwood's KDC C401 10-disc autochanger is a compact unit with 8 times oversampling and can be controlled from a number of Kenwood head units or as a stand alone unit by using the KCA R20M shown on page 125.

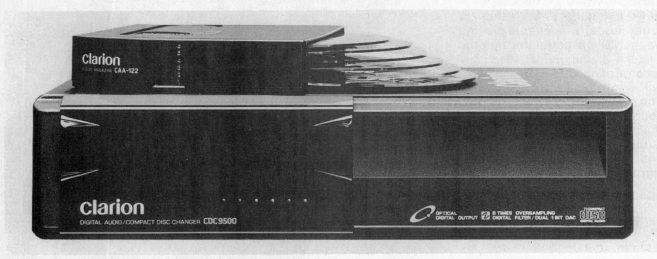

Compare it with the Clarion CDC 9500, 6-disc changer, the same in essence, it also has 8 times oversampling, and its inherent smaller size could be of use when mounting space is tight. As with many changers, the CDC 9500 can be mounted at angles other than 90 degrees - in this case, between 15 – 120 degrees!

At right is the Pioneer 6-disc magazine alongside Alpine's version. Though they look the same, they are incompatible, not least because the discs are inserted in a different manner. The Pioneer magazine has been manufactured in such a way as to be compatible with that company's domestic CD players and so the same magazine can be used in-car or in the home.

Choosing and fitting a CD tuner and autochanger

Mounting an autochanger can be simple or difficult, depending largely on the car you drive. The main points are that you must be able to insert and remove the magazines/discs, it must be very secure to prevent disc skipping and it must be safe from possible harm from people and/or luggage. Saloon cars have an excellent place, just under the rear shelf. Mounted thus in Auto Audio's BMW demonstrator, this 10-disc, Sony autochanger is well out of the way of most luggage and handy for swapping the magazine.

If you have a purpose-made, odds-and-ends cubby hole in your car, then it could be that the autochanger will fit in there. Nakamichi's 10-disc changer is very happy in this Passat, though a special mounting panel had to be made to keep it stable. (Courtesy Audiofile)

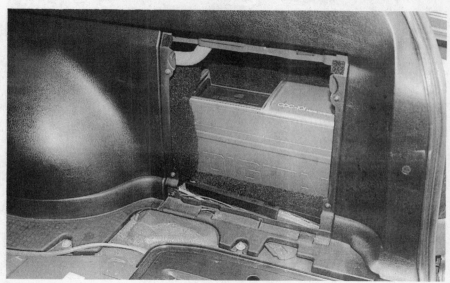

In the Chasseur Jaguar special, the whole of the boot had been retrimmed and adapted to suit the sound system. Note that the crossover unit, at left, is usually covered by a leather trimmed box which also houses a spare magazine for the autochanger. The latter is mounted in a purpose-made box which sits alongside the spare wheel.

Choosing and fitting a CD tuner and autochanger

If you have a suitably shaped glovebox, then popping a CD changer inside is ideal. The Alpine 5952V fits snugly here (there's even enough room for a spare magazine) where changing the magazine is simplicity itself.

THE 'before you start' checklist
The list relating to the fitting of a CD player or CD/tuner applies here with these additions:
10. As well as the above preparation, you need to ensure that you have enough to space to mount an autochanger and get to it to change the magazine.
11. You'll probably need to run wiring from front to rear of the car, much of your time will be taken up with removing and replacing trim and/or carpets and/or seats, so you will need to allow plenty of time.
12. You'll need to allow extra time for cable routeing and the associated trim removal replacement.

For this section, a Clarion CDC 9300, 6-disc autochanger was fitted by Dards Electronics to a Ford Escort. In this particular instance, it was to be linked to the CRX 91R head unit already fitted in the car's DIN aperture. As such, it could be operated directly from it and there was no need to mount a control module. Despite their apparent complexity, most CD autochangers come with far fewer bits and pieces than a radio/cassette deck.

All changers are fitted with securing screws at the factory, to prevent the interior mechanism from bouncing around in transit. You don't have to remove them straight away (in fact, it's probably best to leave them there whilst you are actually fitting it) but you must remove them before you attempt to use the machine. The Clarion machine has six on the underside, as seen here.

Choosing and fitting a CD tuner and autochanger

Another screw, this time on the side of the changer. The CDC 9300 can be mounted vertically or horizontally and the screw should be turned to suit.

As with most hatchbacks, mounting a changer is a little problematical if the practicality is to be retained and you are to avoid the complications of making up special panels to fit it. As a company, Dards Electronics do plenty of the latter, but it's high skilled work and somewhat out of the range of this book. In a saloon, it could be mounted under the rear shelf – easy to get at and mostly out of harm's way. In some cars, it is possible to mount a changer under the passenger seat, though some care has to be taken that it won't be damaged by errant feet. In the hatch is the almost universal choice. Before you even think about drilling holes, CHECK THE POSITIONING OF THE PETROL TANK! Plunging a red hot drill bit into a tank full of highly inflammable petrol fumes is not to be recommended. The position shown here would be OK from a luggage point of view, but it would restrict loading potential if the rear seat were to be folded.

As folding the seat was likely to be a regular occurrence on this particular car, it was decided to mount the changer at the rear, nearside of the hatch on the floor. The multi-hole brackets are designed to fix to the unit in various ways, depending on which way up it is to be mounted.

Choosing and fitting a CD tuner and autochanger

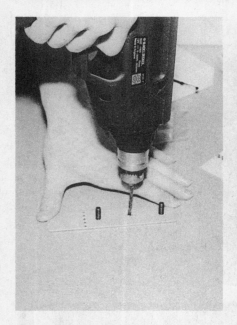

The changer slots onto suitably threaded studs from a permanently fixed bracket which has to be drilled to suit the car in question.

Even though he was some way from the petrol tank, Barry still checked thoroughly what was directly underneath the boot floor. It is still possible for there to be cables, pipes or strengthening sections at the points where you want to drill.

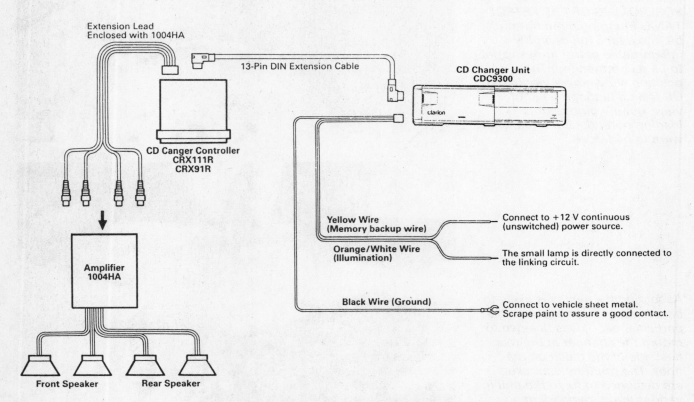

The wiring is very simple, as can be seen from this diagram. Most of the complexities are taken care of by the use of a multi-core cable, culminating...(Courtesy Clarion Shoji (UK) Ltd.)

Choosing and fitting a CD tuner and autochanger

... in a DIN plug which simply plugs into a socket on the end of the charger, with the other end ...

... plugging into a similar socket on the back of the deck. Obviously, extreme care must be taken when routeing this cable (especially in a folding rear seat hatchback) so that it does not become trapped or become a danger to the occupants

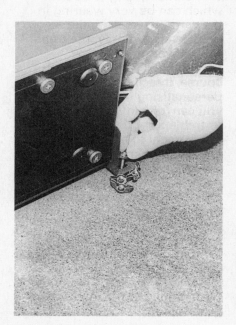

The brackets on the end of the set slot neatly over the threaded studs and the changer is secured by four wing nuts. The secure mounting of the changer is absolutely essential if you are to enjoy crystal clear, unhindered music. The damping mechanisms on any CD player are designed to work best when the changer itself is rock solid in its position.

All that remains to do is to load up 6 discs from your Beatles collection (well that's my choice, anyway!) and switch on the deck at the front. The radio/cassette controls double up to operate the autochanger.

Chapter 10
Choosing and fitting a Graphic Equaliser and DSP

A graphic equaliser is, in essence, a sophisticated tone control. Instead of using just one control for bass frequencies and one for treble, they offer between five and eleven different controllable bands. These bands can take the form of slider controls or, on electronic graphics, LED lights. Whichever you choose, you will be able to see how the machine is set by the 'graphic' display, hence the name, it controls (or equalises) the sound in a manner which is graphic. Technically, a graphic equaliser is a form of preamplifier, that is, it processes the sound signals before they get to an amplifier. Even the simplest, single rotary tone control is a preamplifier. For the hostile, in-car environment, a good graphic is a boon. Some give extra tone control but introduce extra electronic noise into the system. Any vehicle abounds with all kinds of extraneous noise; tyre noise, wind noise, engine noise, rattles and squeaks - it's a wonder you can hear anything else!

If you have ever blown over the top of a milk bottle, you will know that it makes a vaguely musical note. This is caused by the air vibrating inside the bottle and is called resonance. This is what is happening in your car and when you try to add sounds from your stereo system as well, you end up with a real acoustic melange.

The graphic equaliser allows you to 'tune out' some of the more annoying areas of irritation. For example, the volume of air in most cars is quite large and the resultant resonance usually corresponds to middle C on a piano. By using a graphic, any annoying 'booms in this sound area can be compensated for by cutting the frequency response in the 250Hz – 500Hz region.

In addition, it can be used to dial out the top-end 'hiss' on some cassette tapes, or to brighten tapes (by boosting the treble frequencies) which are showing their age. Conversely, some Compact Discs produce a very sibilant, 'bright' sound which can be very wearing in the car. By tweaking the graphic accordingly, the treble can be wound down to acceptable levels. Also, of course, it facilitates a personalising of the sound – you can 'dial in' as much bass or treble as you wish.

It is usual to fit a graphic equaliser when the rest of the system is well sorted, particularly the speakers. Some graphics have an amplifier built-in, though many come simply as a pre-amp unit and require separate amplification. With the former, it is important to check whether your head unit is suitable for such an addition, preferably, before you buy the graphic.

Choosing and fitting a graphic equaliser and DSP

This 9-band, Philips graphic is half-DIN size. In this case, some special panelwork has been made to allow the fitment of the graphic and head unit into the centre console. (Courtesy Dards Electronics Ltd)

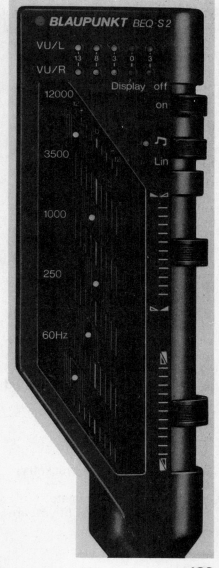

The Alpine 3339 is a multi-choice, 11-band graphic. The separate control/display module can be mounted remotely from the 'works', either on a flexible stalk, a specially made panel or even in the glovebox. Alternatively, Alpine produce a DIN adaptor which facilitates its fitting into a DIN aperture. A major benefit with this unit is that it can be removed from the adaptor, in the same way as a quick-out head unit, for security purposes.

The Blaupunkt BEQ S2 is mounted atop a flexible stalk and can be positioned where best suits the driver. When the car is left unattended, bending the stalk right down means that the control unit can be positioned out of sight of prying eyes. (Courtesy Robert Bosch Ltd.)

133

Choosing and fitting a graphic equaliser and DSP

The 'before you start' checklist

1. Is your head unit designed to be linked to a graphic equaliser?
2. Make sure that the unit you are fitting is complete. Compare what you have in the box to with manufacturer's list.
3. Ensure that the graphic will fit where you want it to. Is it DIN size? Not all graphics are, and if not, you'll need a suitable mounting place under the dash. Remember that there has to be room to route the wiring neatly, safely and correctly. The fitting of brackets and adaptors takes time so allow for it in your planning.
4. Read the instructions and make sure that you fully understand them before you start. Are you sure how the graphic is powered? Some have their own power feed, others take their cue from the head unit.
5. Have you all the terminals/wiring required? You may need to change connectors on either graphic or head unit, make sure you have enough.
6. Spare fuses – 3 just in case.
7. Make sure you have enough time, graphics can be complex to wire and mount. Give yourself plenty of leeway and don't rush the job.
8. Make sure that your speakers are up to the job. Fitting a high power graphic and running it through cheap, wide range speakers is a waste of time and money.
9. Tools for the job. Fitting a graphic equaliser will typically require:
Medium/small crosshead/slotted screwdrivers.
Crimper
12v test light
Torch
Pliers
Electric drill/bits (for bracketry)
Insulating tape
Spare fuses/terminals etc.
Soldering iron/solder (if you are proficient)

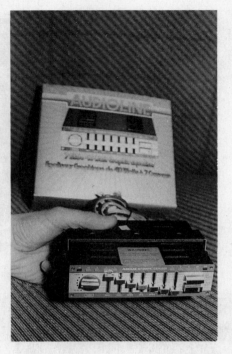

This Audioline 307C is typical of the type of graphic fitted to many systems. It is a non-DIN size unit designed to be mounted on a bracket wherever suitable.

The key features of the set are shown in this diagram. The LED lights at right and left show the amount of power being used through each channel and a front to rear fader is built-in. The LED indicators can be switched off at night so as not to cause a distraction to driving. (Courtesy Harry Moss International)

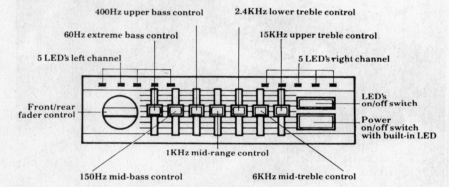

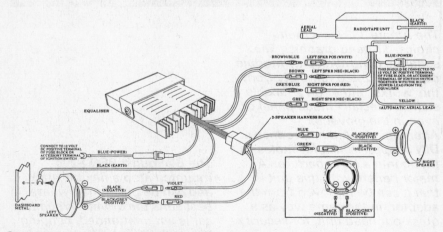

Spend some time and read the instructions until you thoroughly understand them; it is very easy to mix the wiring. This diagram shows how the unit would be wired for two-speaker systems whereas...(Courtesy Harry Moss International)

Choosing and fitting a graphic equaliser and DSP

... it could be wired for a four-speaker set up. Once having made the choice, the wiring is made simpler by the fact that...(Courtesy Harry Moss International).

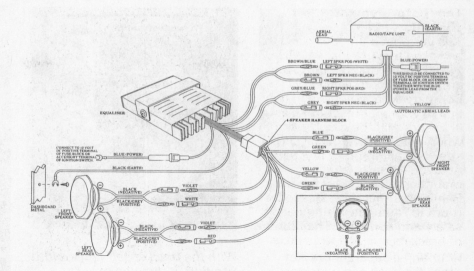

... the 307C comes with two pre-wired sockets – one for stereo systems and the other for 4 speakers.

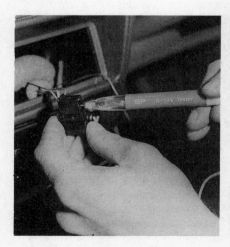

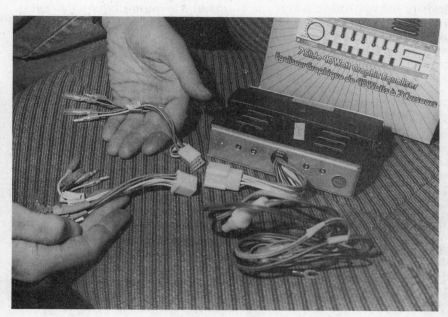

With the head unit removed and power carrying leads checked out, the battery earth terminal was removed.

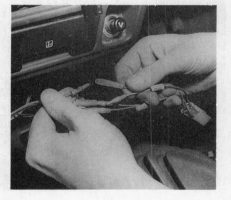

As is often the case, it was necessary to adapt the terminals. In this case, we chose to alter the vehicle speaker terminals to suit the bullet connectors on the Audioline equipment. It is vital to maintain speaker polarity to avoid phasing. The best way to achieve this is to use male/female terminals for each speaker in such a way as to mimic the terminals already fitted to the graphic.

135

Choosing and fitting a graphic equaliser and DSP

Mounting a separate bracket under the dash can be tricky - it tends to move around when you are trying to mark the hole positions. A good idea is to use a double sided 'sticky'...

With the bracket self-tapped into position, it remains only to insert the multi-plug in the rear of the graphic and...

Mounted here, the unit can be operated by both driver and passenger. Whilst it is a little low for the driver, unlike a head unit, it is not the kind of item likely to need constant attention.

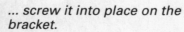

... screw it into place on the bracket.

... to secure the bracket temporarily in place. Not only does this make hole-marking easier, it also allows you to step back and see if it looks satisfactory where you have positioned it. If you're happy, mark the positions through the holes in the bracket, remove it and drill the holes. Make sure that there is nothing expensive or dangerous to drill into, though.

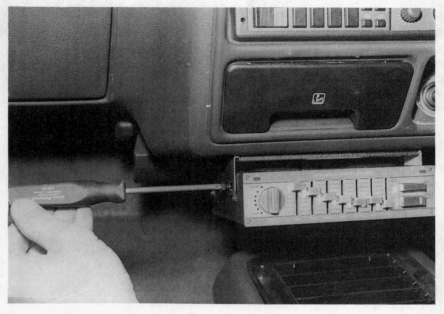

Choosing and fitting a graphic equaliser and DSP

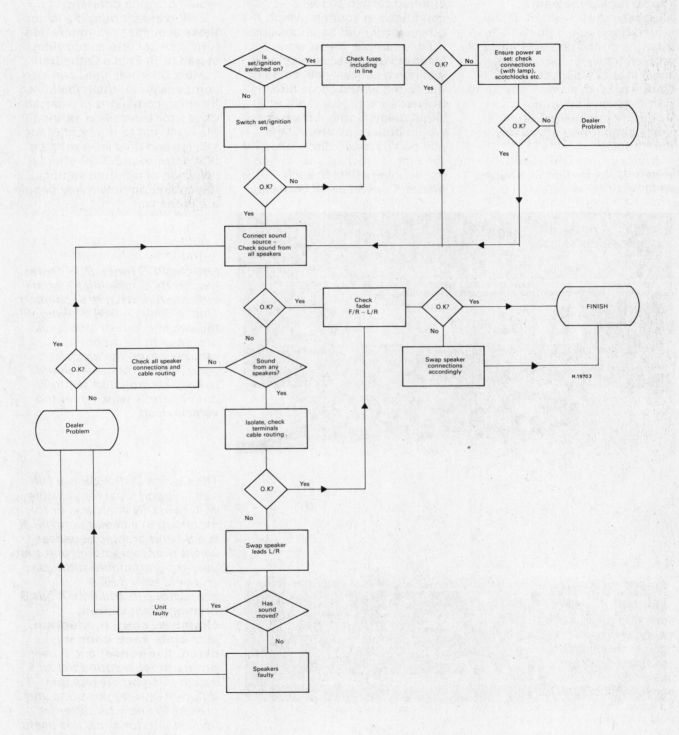

Choosing and fitting a graphic equaliser and DSP

DSP

DSP refers not to some psychic ability, but to a natural development of the graphic equaliser, the Digital Signal Processor. As the name suggests, the advent of digital sound has made it possible to alter the sound characteristics of a given track, digitally. Some machines can perform little 'tricks' as well, such as speeding up or slowing down the music, or removing the lead singer's voice altogether - effectively in-car Karioke!

Basically, the DSP reads the musical information as a series of lumps of computer information. Once it has digested them, it can be programmed to use them in different ways. The main function of any DSP unit is to emulate certain acoustic conditions in your car which otherwise would be impossible.

For example, if you were listening to someone playing a violin in a concert hall, some of the sound would come directly to your ears, though with a slight delay. Some, however, would bounce all around the hall and be much diminished by the time it reached you, hence the echoing effect in such places. Conversely, if you were listening to John Dankworth in a typical jazz club (where hundreds of people cram themselves into the smallest space available!) the acoustics would be quite different.

DSP makes it possible to copy those acoustic parameters and turn your car into a recording studio or St Paul's Cathedral!

Most DSP machines can also compensate for the unbalanced listening conditions in a car; the driver, for example is sitting to the right and to the front of the vehicle and thus is hearing an off-centre sound. DSP offers a selection of different settings, depending on how many people are in the car.

Like the RDS tuner, DSP scores heavily by displaying a visual indication of what is happening. This is Sanyo's DSP 01 where the display/control module is designed to be mounted remotely from the electronic control box. The latter would usually be mounted securely out of harm's way, under the vehicle dash.

The Clarion DSP 959E is a DIN size unit and ideal for vehicles with twin DIN apertures or for mounting in a centre console. It is a 9-band graphic equaliser, with 4 preprogrammed and four user-programmable settings and gold plated RCA connectors. In addition, it has 8 preprogrammed (**hall, chamber, church, stadium, jazz club, rock concert, disco, livehouse**) and 8 user-programmable sound fields, 5 sound imaging presets for different seating positions and 3 presets to allow for different vehicle interior sizes. It is useful to plunder Clarion's own descriptions for details of the machine and DSP in general.

Choosing and fitting a graphic equaliser and DSP

INITIAL DELAY
This is the delay time between the direct sound being emitted and the first early reflected sound being heard. It represents the distance between the sound source and the closest reflective walls.

ROOM SIZE
This is the time interval between the early reflected sounds. It represents the size of the sound field space, the higher the value, the larger the room seems.

LIVENESS
This is the degree of attenuation of the early reflected sound and represents the way the sound reverberates in the sound field space. In **disco** mode, the reverberations are set high, while in **jazz club** mode they are low.

REVERBERATION TIME
This is the length of reverberations in the sound field. The higher the value, the longer the reverberations, producing an echo-like effect. This value is set high for **church** mode, to recreate the effect of high ceilings and bare stone walls.

HIGH
This is the ratio of treble sound in the reverberations. The higher the value, the higher the sound of the resonance. Clarion claim that all this technology adds up to a total of 770 different sound fields available from this single machine! Whatever the technical aspects of DSP, the resultant sounds are staggering and have to be heard to be believed. There is little doubt that, given a price decrease similar to that of CD players, DSPs will take over from conventional graphic equalisers.

Chapter 11
Choosing and fitting a Power Amplifier

In this chapter we are dealing with power amplifiers as opposed to preamplifiers. The difference is simple; a preamplifier actually alters the signal received before it gets to the power amp. Therefore, any form of tone control or graphic equaliser is, effectively, a pre-amplifier. A power amp simply boosts the signal it receives. They are available in many shapes and sizes and almost any output you require, from the sublime to the ridiculous.

An amplifier of any description will always need to be kept cool; the more power produced, the cooler it will need to be. This is usually achieved by means of cooling fins on the outer casing of the amp (rather like the fins on a motorcycle cylinder head) and it follows that for best efficiency, an amp should always be placed where cooling air can circulate.

One of the most popular mounting places is under the passenger seat. This is a fairly cool place and makes wiring to the deck relatively simple. It is quite possible to mount an amp in the boot or hatch of your car. However, it means that you have a fair way to run those wires which may need to be extended (for which you are pointed in the direction of proprietary extension leads) and it can leave the amplifier open to damage by carelessly loaded luggage. Another good mounting place is actually beneath the dashboard. This is not always easy, however, and although many dealers fit amps there, they have the skilled staff to make-up the mounting panels and they also have the knowledge to ascertain whether or not the amp will be cool enough.

BUYING
Before you even think about buying a power amplifier, consider the rest of your system. Some head units are not suitable for external amplification, check with your supplier before you start spending your money. If it is suitable, then you should be brutally honest with yourself; is it worth amplifying? As a reasonable basis, you'll need a good quality head unit, with metal tape facility and some form of noise reduction on a cassette deck. (All but the very cheapest CD players are worth adding an amp to.) If you have A poor head unit now, adding an amplifier will leave you with the same poor sound, but louder. Think about your speakers and remember the rule, add 50% as a margin. So, if your amp produces 20W max per channel (see below), your speakers should be capable of handling at least 30W max. Always upgrade your speakers before your amplification, otherwise you'll not hear your amps at their best and, in certain circumstances, you could actually ruin the speakers altogether.

HOW MANY CHANNELS
Amplifiers generally come in

Choosing and fitting a power amplifier

2/4/5 or 6 channel formats and which you should choose depends on a number of factors, including; your system, personal choice, depth of wallet and physical space available. In many cases, it is the latter which takes precedence. In my own car, a single 6-channel amp would have been an excellent solution to providing the motive power, but it was far too large to fit under the seat (or anywhere else, for that matter) and so I ended up running 3, 2-channel amplifiers to achieve the same result. Clearly, this created extra work in mounting and wiring the amplification and it also provided lots more opportunities to get it wrong.

When you're looking at a head unit with an integral amplifier, make sure that you read what you see; 20W per channel, could be either 80W total or 40W total, depending on how many channels are quoted. Some machines are designed to be 2-channels and others 4.

TOTAL HARMONIC DISTORTION
Thankfully abbreviated to THD, this is a measure of how much noise an amplifier generates in order to produce a given power output. It is similar to comparing, say a Citroen 2CV driving at 80 Mph and a Mercedes 600 driving at the same speed, one would be rather louder than the other. For example, an amp may produce 100W but with a THD of 10%, a high output, but an equally high noise level. Look for figures at 1% THD or less. The THD figure is almost directly linked to the price ticket, you get what you pay for.

FREQUENCY RESPONSE
The frequency response is a measure of the highest and lowest notes your amplifier is capable of reproducing. The figures quoted will be in either Hertz (Hz) or kilohertz (thousands of hertz or kHz); the lower the figure, the lower the note. As a guide, sounds below 200 Hz come into the sub-bass category and anything much above 22,000 Hz (22 kHz) is beyond normal human hearing. Obviously, the wider the frequency response, the better sound you're likely to hear, as long as you have speakers capable of reproducing that sound.

SIGNAL TO NOISE RATIO
The signal to noise ratio is denoted in dB (decibels) and is a measurement of how efficient the amplifier is in delivering to your ears the pure sound from the tape deck as opposed to the internal noise generated in the process of amplification. The lower the figure, the better, where you can regard 80 or 90 dB very good - but not cheap.

UPGRADING
You could use some ingenuity when you're upgrading yours system, so that you don't lose money on your existing power amps. Say, for example, you have a 2-channel amp running the rear speakers, with the front running through the integral amp. You could add a 4-channel amplifier to power the front and rear speakers, utilising the original, 2-channel, unit to power a sub-woofer speaker or tube.

THE POWER RATING
Amplifier power is usually given in two ways - maximum (music) or continuous (rated). The former is a measure of how much power the amp can produce for a very short period of time. The latter is a measure of the power that the amp can produce over a much longer period of time and is a much more realistic figure. The difference between the two is a guide to the amplifier's efficiency; prefer, for example, an amp producing 100W max/80W continuous to one producing 100W max and 60W continuous. When comparing amps, make sure that you are comparing like with like. Another major problem is that some amplifier manufactures (like some car manufacturers) get carried away when they're rating their amps.

As such, you could find that two amps ostensibly giving the same power output, could sound markedly different. By all means, trust in a manufacturer's good name, but ignore the writing on the case and trust what your ears are telling you; always make sure that you listen before you buy.

Split between 4 speakers (i.e., no sub-woofing) 100W max of good quality amplification makes for a nice sounding system, given that everything else is up to scratch. You'll need that sort of output to make the best of a CD based system, with its extra dynamic range. Personally, I feel that around 200W RMS for the four speakers (or components) with 100W RMS for the sub-woofing, is about ideal for a system that is not intended for public showing. There certainly won't be any danger of not hearing it, regardless of engine/wind/tyre noise. A correctly installed system, using good quality equipment, running this kind of power will rank with the best of them, and I speak from experience. Nevertheless, if you want more power, there's plenty on offer, but I would suggest that you listen first.

Remember that the more power an amplifier produces, the larger the drain on your car's charging system. The manufacturer fits any car with an alternator to suit an average set of parameters. 500W of music power is not average, and you could soon bring your battery to its knees, even with the engine running, especially if your battery is not in its prime.

Choosing and fitting a power amplifier

For very high powered systems, it will be necessary to fit a separate charging system and battery altogether. However, this is heavily into the realms of the professional.

SAFETY

You must make sure that your amplifier is fitted safely. It should be mounted securely, out of harm's way and where it is away from any source of water for example, dripping rain from a bootlid. Similarly, use the correct size cable - a powerful amplifier could cause thin cable to overheat, with the obvious, unpleasant consequences.

When listening to your newly amplified system, it's quite possible to get carried away with the volume setting, more so as your ears get used to ever higher volumes. Make it your policy to turn the volume down (or off) every so often, not only to give your ears a rest, but also to make sure that there are no nasty noises coming from your car!

On a more serious note, your hearing is both precious and delicate, and it is surprisingly easy to damage it permanently by injudicious use of the volume control. There are cases where the inner ear has been damaged by too much noise, causing not only hearing but also balance problems as well.

Also seen at the Sound-Off, was the Philips Monster van. Designed to be an 'ultimate', the amps on board (of which those seen here are just a few) produce well over 5000 watts!

Who said it's difficult to fit amplifiers in a sports cars? The owner of this Porsche 911, seen at a National UK Sound-Off championship, thought differently. Trimming and fitting to this high standard takes no small amount of time, money and skill.

Choosing and fitting a power amplifier

From lots of amps in large vehicles, to the other extreme. This Kenwood-equipped mini was a special project and shows what can be done in a small car. Again, much specialised trimming is in evidence, with most of the rear of the car being converted to house the ICE gear. But who needs boot space anyway?

There's some semblance of practicality, here, suitably so in Vauxhall's practical coupe, the Calibra. A sub-woofer/speaker cabinet has been constructed to sit between the rear suspension mountings, the rear of which comes in handy for mounting extra amplification and a 10-disc CD changer. Though luggage capacity has been reduced (and the rear seat will not now fold forward) there is still quite a reasonable, flat loading space available. (Courtesy Leicester Car Radio)

In some respects, saloons offer advantages in amplifier mounting, simply because most do not have a folding rear seat. As such, amplifiers can be positioned as shown here, which leaves plenty of luggage space available and they are well out of the way and unlikely to get damaged.

Choosing and fitting a power amplifier

Professional installers soon get to know the ins and outs of different cars and in particular, where there are little cubby holes just waiting for an amplifier. There's one such space in the nearside rear of the VW Passat estate and—
(Courtesy Audiofile)

—a similar one under the offside flooring. These two Nakamichi amps fit nicely in there, though a special panel has had to be made and fitted to keep them secure. (Courtesy Audiofile)

The 'before you start' checklist
1. Make sure that what should be in the box, is in the box.
2. Ensure that the amplifier will physically fit where you want it to go and that it is suitable for your head unit/speakers.
3. Make sure that the amplifier will be kept cool.
4. Make sure that you fully understand the instructions - before you start.
5. Check on your stock of terminals, spare wire etc. Have you got the made-up leads you required (if any)?
6. Make sure you have a spare fuse or two, just in case.
7. Give yourself enough time to do the job, with a little allowed for fault-finding.
8. Tools for the job. A typical fitment will require;
A crimper (and/or soldering iron)
12v test lamp
Torch
Electric drill/bits
Screwdrivers
Electrician's tape
Craft knife

For this fitting section, I watched the fitting of a Philips DAP 300 amplifier (specification at the end of this chapter) at Dards Electronics of Milton Keynes, to a system as part of an upgrade.

The system in question was originally a Philips DC980R CD/tuner which played its integral 50W through four custom-fit speakers. The object was to uprate the whole system, a piece at a time, money being a restricting factor, meaning that work had to be carried out in stages. The rear speakers had already been uprated to Philips PRS 602 components (see chapter 3).The head unit was equipped with 4 pre-out sockets, meaning that front and/or rear channels could be played through a separate amplifier (or two). In this case, the rear speakers were to be powered by the new amp leaving the front speakers powered by the integral amplifier.

Though the maximum power from the separate amplifier was distinctly higher than this, it was offset by two factors; a) the two,

Choosing and fitting a power amplifier

6" woofers at the rear need some grunt to enable them to shift the right amount of air and b) a further uprate was planned at a later date for the front, which would go even further toward restoring the status quo. By turning down the gain on the power amp, it was possible to create a reasonable balance between front and rear.

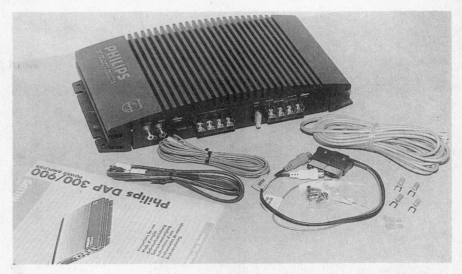

This is where we start. The amplifier out of the box, standing alongside the various wires and connectors supplied, checking them off against the parts list. Some time with the instruction manual was a necessity to ensure correct fitment. As is common on good amplifiers, the terminals are gold plated for better signal transmission. There are twin 20 amp fuses and twin RCA sockets for the L/R input from the set.

The time before fitting is the time to get familiar with the various settings and controls; when it's mounted on the floor of your car makes life more awkward than necessary. Here is the screwdriver- operated gain control together with filter and input mode selections.

The car was a hatchback, which gave little choice of mounting position, if luggage capacity was to be retained. Under the front passenger seat was the choice. As the power for a separate power amplifier comes direct from the battery, it was necessary to drill a hole in the bulkhead (there were no suitable ones there, unfortunately). A grommet was, of course, fitted to ensure that the wiring did not chafe.

Choosing and fitting a power amplifier

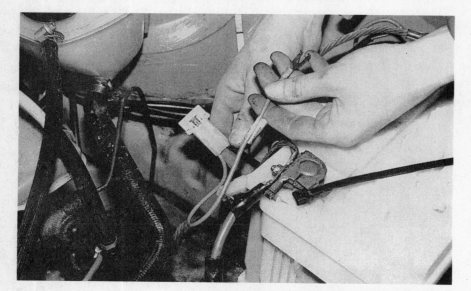

With the earth terminal removed, the amplifier power lead was connected to the positive terminal. Note the in-line fuse close to the terminal. It is important that any such fuse rating should not be altered, it is as the manufacturers intended and should be left that way.

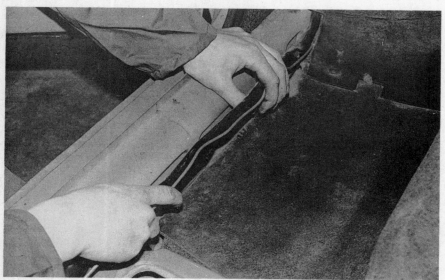

The front seat was removed (sounds drastic, but it's only 4 or 5 bolts in most cars) and the carpet from front to rear pulled well back. This allowed the power cable to be routed down the side of the car, under the carpet. Care had to be taken not to route the cable where it could be trapped or chafed.

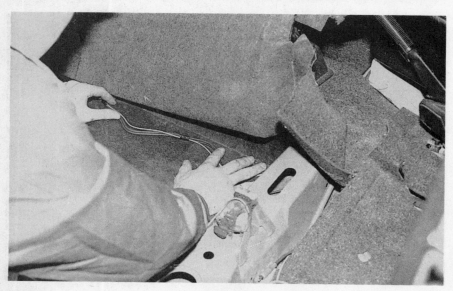

Conversely, the wiring to the set was taken along the centre tunnel and up the centre console.

Choosing and fitting a power amplifier

The wiring was not difficult, using the loom and terminals provided, though great care was taken to ensure that the terminals were secure.

This diagram shows what goes where, and there is little here that the average DIY fitter could not handle.

Key
1 INPUT SOCKETS
2 7-PIN ADAPTOR
3 RCA CABLE/PLUGS
4 POWER CABLE
5 EARTH CABLE
6 REMOTE ON/OFF
7 SPEAKER TERMINAL
9 SECOND REMOTE

The amplifier is wired using the remote cable, so that it is switched on and off automatically with the set. The second remote cable is used for other car stereo components, such as a CD changer. (Courtesy Philips Car Stereo)

The amplifier has to be earthed, the nearer the better. A hole was drilled in the cross strut, as shown here and—

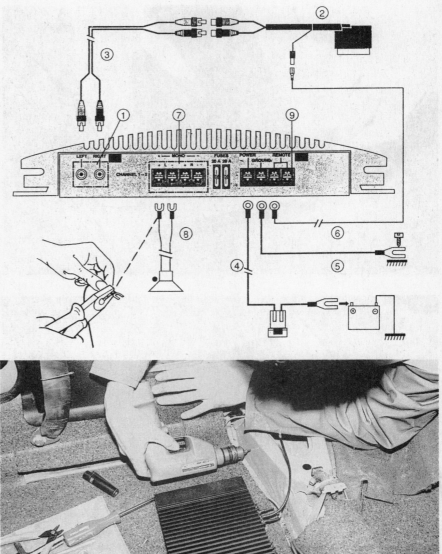

147

Choosing and fitting a power amplifier

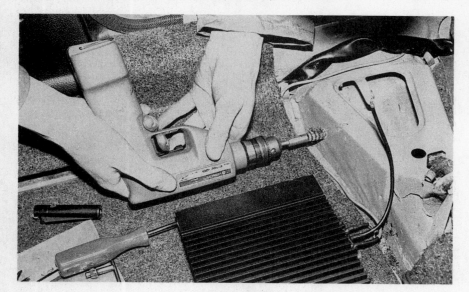

—then, using a small wire brush in the cordless drill, it was relieved of its paint to ensure a good metal to metal contact. Don't forget that poor earths account for an enormous number of ICE problems.

As you can see here, the twin RCA leads carrying the sound from two rear channels, simply plug into the amp at the front. Note that the amplifier was positioned in such a way that all the connections were facing the front of the car, out of the way of 'size '9s'. Though it looks as if the amp is being screwed into the vehicle floor, it's actually being self-tapped into the carpet and underfelt. Because the unit is reasonably sized and positioned well forward, it's unlikely to move, even if knocked by a rear seat passengers feet.

It's almost a shame to cover it up, but the front seat passenger wasn't too happy about sitting in the back! Note how the wiring has been loomed together where possible (not power and speaker leads) and that the installation looks good, even though it is not going to be seen. This is a professional trait but one you'd do well to emulate.

Choosing and fitting a power amplifier

Having got to this stage, it remained only to make sure that the settings were correct. Bridged mode (linking the two stereo channels into one mono channel) was not required. The gain setting is, effectively, a volume control, and is useful in a situation like this, where it can be used to tone the rear part of the system down, to prevent it from being too unbalanced. Ideally, the sound stage (i.e. where the sound appears to be 'positioned' in the cabin) should be in the middle of the car.

A multi-amp system should sound good, but there are certain practical considerations to be borne in mind ...

Specification
Philips DAP 300 power amplifier
Max power output   2 x 80W or 1 x 150W
Power output at   0.08% THD   2 x 55W or 1 x 110W
Freqency Response   5Hz - 70,000 Hz
Signal/Noise Ratio   105 dB
Dimensions   175 x 305 x 62

Chapter 12
In-Car Telephones

In-car telephones are not actually a new invention, the basic system was available over 30 years ago. However, as with other ICE equipment, it is the miniaturisation of electrical components which has made it practical to bring car phones to the mass market.In some respects, the progress of the in-car phone since the mid 80s can be likened to that of the Compact Disc.Initially, the carphone was an extremely expensive item, suitable for pop-stars and oil magnates.But the demand was such that the price of equipment (and lately, air time) plummeted and they have passed quickly and easily into everyday life.No longer are they the prerogative of the large business and, whilst most are still business-linked to some extent, their use is spreading more and more into the private sector.Certainly, they are no bad thing to have in the car if you are a lady travelling long distances alone, for obvious reasons.The AA were the first motoring organisation to utilise specific in-car phones (capable only of calling them in an emergency) for just this reason.For some, the pure convenience of the cellular phone will be enough - anyone who has searched long and hard for an unvandalised telephone kiosk will vouch for that!

HOW DO THEY WORK?
If you have read in earlier chapters relating to radio reception and radio aerials, you may be puzzling as to how it is possible to install a usable telephone in the hostile, mobile environment of a car.The answer lies in the names of the phones themselves, prefixed, as they are, by the word 'cellular'.

For the purpose of telephone signals, the air-time suppliers have divided the country into small 'cells', looking, in schematic format, like a bee's honeycomb.When a signal is transmitted from a phone, it only has to travel a short distance to a transmitting station, on the edge of the cell it is in.

From there, it is relayed to a computerised exchange which decodes the information to decide whether you have dialled a domestic number or another car phone.In the case of the former, it passes the call along the normal land-line route.If the latter, then the signal is sent to the transmitting station in the cell that the receiving number happens to be in.When your car is moving, you may travel out of one cell into another.When this happens, the transceiver will sense a deterioration in the signal and automatically search for a better one.This takes just milliseconds and will not normally be heard by the user.The rules governing FM radio-signals apply to in-car phones and this means that certain factors will prevent the signals getting through.Entering tunnels or passing under bridges will cause interruption and so will driving through a range of hills.

In-car telephones

BUYING CHECKLIST
The in-car phone market expanded at a phenomenal rate at the end of the 80s and allowed the ingress of a large number of 'cowboy' fitters and suppliers, all anxious to take their cut.The result was that a lot were badly installed (to the point of being dangerous) and a lot of customers got their fingers burnt as the 'smart operators' moved on to their next 'kill'.The recession and a tightening up by the major carphone suppliers has meant that those companies still in the carphone business have had to become (if they weren't already) lean and competitive, which has led to an improvement in standards.Nevertheless, you should exercise great caution when buying a phone - the sharks are still out there, looking for innocent bathers in the waters of in-car communications.

BUYING NEW
Buying a new car phone is relatively simple - as long as you go to a reputable dealer.How do you know which is which?Take a look at the phone suppliers he deals with.If he is 'on-board' with the major names, then you can rest assured that those companies will have carefully vetted the dealer in order to ensure that their own reputation is not sullied by a third party.Buying a new carphone from someone's car boot ("...discontinued line guv...") or from a shady, backstreet lock-up is a recipe for disaster - make it someone else's'.

BUYING USED
Buying a used phone is fraught with danger for the uninformed.Your problems start in deciding whether or not the phone in question actually works, not easy for the man-in-the-street.

If the phone you are buying is still operating on the sellers' number, you will soon be able to see if it works correctly, at that time, at least.Short of testing it for a week or so, you won't know if there is any major and/or intermittent fault.If, however, the seller has already transferred his number to a new phone, there is no way you can check whether or not it works.

But the electrical function of the unit is only the tip of the iceberg.Every telephone has two numbers - the number you dial and a serial number.This is sometimes to be found on an official sticker, though not always.It will be on the seller's original air-time agreement, as long as the phone has not been swapped previously.It is also electronically encoded in the phone's electrical innards, not exactly handy for the layman.The system works like this. When the air-time supplier connects the phone into its system, it registers both numbers.When a call is made, computer checks are made to ensure that they tally.So far so good.When the seller changes his phone, he informs the air-time supplier of the new serial number and this is then logged against his existing phone number.As long as the seller is honest, that is that.But if the seller has run-up a bad debt with the air-time supplier, then the company will black list the serial number.If you buy a blacklisted phone, regardless of your innocence, good faith and everything else, you will not be allowed to use it, for the number will come up as 'bad', whichever air-time supplier you choose.

If this sounds complex, it is!What you need to know is that the phone's serial number is 'free' and that the serial number on the casing (if any) relates to the actual phone itself.If you can obtain the serial number, the any air-time supplier will check it for you.However, you should seek to take the phone itself to be thoroughly checked over by the dealer who will be able to check its workings as well as its legality.At the end of the day, if you have any doubts as to whether the used phone you are looking to buy is absolutely above board, turn around and walk away; there are so many phones to choose from, it's pointless to take a risk.

ANTI-LOCK BRAKES
The use of antl-lock braking systems is becoming increasingly popular, and quite right too.However, it has been known for some phones to interfere with some systems.The advice, should your car be so equipped is simple: DO NOT attempt to fit a car phone without consulting a specialist and/or the car manufacturer, your life could depend on it.

ENGINE MANAGEMENT SYSTEMS
In order to make cars more efficient, many manufacturers rely on complex electronic engine management systems.This too is a tricky area when it comes to in-car phone installation.As with anti-lock braking, one of the main concerns is the fact the a car phone transmits signals which could, conceivably, interfere with the car's own electronics.Mid 80s (and onward) Jaguars are legend for their complexity in this direction and the ease with which you can inadvertently create massive problems.Again, the advice is to make sure that you know exactly what you (or your fitting dealer) are doing before you starting snipping cables.Check with the manufacturer first.

POWER
The power of in-car phones is 'Class' rated, from Class I to

In-car telephones

Class IV.Generally, the RF power output is a foregone conclusion, with the small hand-held portable units being Class IV.This is due to the fact that micro-waves are produced in signal transmission and to have to many of these things buzzing around next to your ear could be dangerous.Thus, the power is deliberately limited.Most permanently mounted in-car units (mobiles) are rated as class II, 2.8 watts.Class I phones are very powerful, in fact, so powerful that they could interfere with other car phones, causing all manner of problems.Unless you have a hearing deficiency, you are not likely to find a problem with any of the standard in-car power ratings. For in-car use, Class II is generally just right.Consider the features offered on the phone you are about to purchase.Although buying second-hand is cheaper, in-car phone technology is proceeding apace with the rest of the ICE industry and ever more features are being provided with each new model, often at a lower price.

AIR TIME

Air-time is the time spent actually using the phone.Unlike conventional domestic phones, British Telecom do not have a monopoly.It is, in fact a duopoly, Cellnet and Vodaphone being the two companies who can supply air-time.However, they cannot do so direct and thus we come to air-time suppliers; in order to have your phone operational, you must enter into an agreement with one of these.The agreements are usually a minimum of 12 months.Unless you're being offered a particularly good deal, then try to avoid being tied-up for a longer period than this in order to keep your options open.

Your air-time payments consist of a monthly rental fee and then the costs of the calls.Like land-lines, the rental fee is payable whether you use the phone or not.Like the rental, the cost of calls varies from company to company.The rate per minute is quite a lot higher than for domestic lines and there are varying methods of charging.The 'normal' method is to make a minimum charge of one minute, regardless of how many seconds are actually used and thereafter in 30 second intervals.Some companies work solely in 60 second intervals, a difference which, though it may not seem much, can work out to be very expensive over a period of time.Shop around and take your time is the answer; there are a good many air-time retailers, all of whom want your business.

Choosing between Vodaphone and Cellnet is a matter of personal choice.Chat to someone local to you and/or who uses his phone in the same way that you would.Is he getting good service?Does he find that there are many areas which do not get good reception?

SUMMARY

1.Any car can be fitted with a telephone, given that it has a suitable power source (12v, negative earth) and an aerial.
2.Portable phones will work in-car but if you are using the standard aerial, you will find reception more than a little sporadic.
3.You can fit your own phone and/or transfer it to another car without having to inform anyone.
4.In-car phones have the same facilities as their domestic counterparts (world-wide STD dialling etc.) and more besides.
5.Regardless of who fits your phone, it will not work until it is officially connected into the network of your choice (Cellnet or Vodaphone) by an official air-time supplier.This has to be done by the supplier.
6.Exercise extreme caution when buying a second-hand phone.Check-out the comments made earlier.The 'bargain-price' phone you buy from a bloke in a pub, could turn out to be one of your costlier mistakes.
7.Fitting a car-phone yourself is possible.If you're not totally competent, though, take it to a reputable dealer.It is surprisingly easy to ruin a phone and they are not cheap to repair.
8.Whether you're fitting yourself, or going professional, get hold of a copy of the DTi Code of Practice for the installation of mobile radio equipment, ref MPT 1362.It is a particularly concise and readable document, especially for one Government produced, and will prove interesting reading; it tells exactly how your phone should be fitted.

SAFETY:

Clause 49a of the Highway Code states that you should not use a hand-held microphone or telephone handset whilst the vehicle is in motion, except in an emergency.Also, you should never stop on the hard shoulder to make or receive a call, no matter how urgent.Obviously, if your car has actually broken down, then making a call from your phone inside the car would be a safety aid, as it would not necessitate walking alongside the carriageway to an emergency telephone.

INSURANCE

Having got your phone installed, you should tell your insurers straight away.Some companies may specifically exclude them, whilst some may add them to your policy for a small extra fee.If you have problems, there are many companies who specialise in insuring carphones.Whichever, you would be well advised to

In-car telephones

cover the possibility of theft or damage.

GLOSSARY
Electronic Lock
The Panasonic phone in this section follows the usual method of phone-locking, in that it is a two-stage affair. The phone can be barred from making certain types of calls, usually overseas calls, or it can be locked altogether, preventing anyone from making any calls at all. However, regardless of the lock selected, it will still receive incoming calls.

Call monitor
Allows the caller to monitor the progress of the call (e.g. dial tone, engaged tone etc.) without lifting the handset until the phone is answered.

Displays our numbers
A feature where the phone displays its own number. Useful where a car is used by more than one driver or where the owner has a bad memory!

DTMF Signalling
Allows access to data systems using DTMF signalling.

Elapsed call time
A facility which enables the call duration to be monitored and displayed.

Illuminated display
As the name suggests, this relates to backlighting of both keypad and LCD display.

Key confirmation
When a key is pressed, a small audible 'beep' confirms that the instruction has been received.

Last digit clear
Wrongly entered numbers can be cleared from the display one at a time or as a complete sequence.

Last number redial
By pressing one button, the last number dialled will be called again.

Memory scrolling
The ability to 'scroll' through the memory banks to find a specific stored number or a blank space in which to store a new one.

Mute control
The ability to switch off the in-car microphone (handset or remote) to allow a private conversation whilst taking a call.

Noise cancelling microphone
A hands-free microphone which automatically cancels out most of the background noise inherent in an in-car environment.

On-hook dialling
Enables the dialling of a number whilst the handset is still on the cradle. It is quite possible to fit your own in-car phone, but you must be very competent and patient. What experts do in four hours may well take you a whole Saturday and some of Sunday as well! Remember that the wiring usually has to be routed from one end of the car to the other and this involves a lot of time spent removing and replacing trim panels. It is also not a job which can be 'bodged' lightly. Installation of your radio/cassette deck will inconvenience no-one but you and your unfortunate passengers. However, because the carphone sends as well as receives signals, you can cause problems for other carphone users if you get it wrong, and sometimes it is possible for a badly installed cellular unit to interfere with the emergency services. Study the fitting section in this chapter; if you can't match the attention to detail shown here, the take it to a dealer. For this fitting section, I watched the fitting of a Panasonic 'E' series transportable phone at Servicom's base in Redditch. Boss, Martin Scholes arranged it and the fitter was Chris Matthews; my thanks to them both.

A transportable phone combines the best features of a dedicated unit with the portability of a fully mobile phone and the 'E' series Panasonic is certainly one of the best around. Rather than having the transceiver (the electronic box which sorts out incoming and outgoing signals) mounted permanently in the car, it is held in a bracket, usually mounted in the boot or hatch. Whilst there, it is powered from the car's electrical system (much more reliable than batteries) and uses the aerial perched atop the vehicle roof. When it is required outside the car, or in someone else's car, the handset is unplugged from the cradle and clipped into the transceiver which, in turn, can be unplugged from its mounting bracket.

A built-in nicad battery then powers the phone (it receives a trickle charge from the car battery when the transceiver is in its bracket) and if you take it into a car with a cigar lighter socket, it can be powered from that. For true versatility, you can't do better.

In-car telephones

We start with the phone unit itself. It can be installed as a dedicated carphone without the transportable option, which comes separately. Potential purchasers could buy the phone and add the rest at a later date if cash were tight. The car-mount kit includes cable, bracket for transceiver, cradle for handset, microphone and speaker.

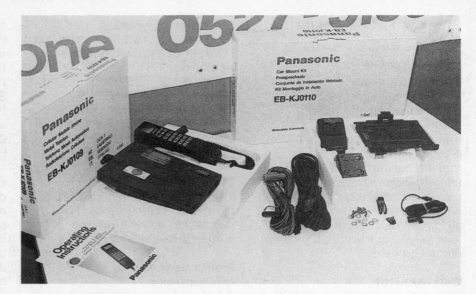

The transportable kit includes a transceiver adaptor, case, mains battery charger, cigar lighter adaptor and nicad battery, the latter of which is —

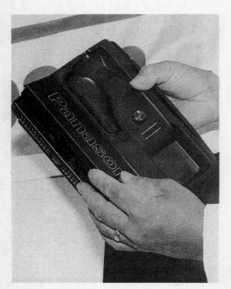

—slotted into adaptor case, which in turn—

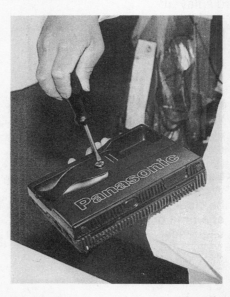

—screws onto the transceiver, whilst the phone handset—

In-car telephones

—is taken from its standard resting place inside the car and plugged directly into the transceiver. Finally—

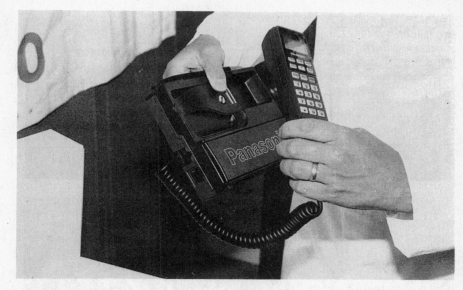

—the stubby, but powerful, aerial slots in and that's it, you're mobile. The whole conversion process takes around a minute.

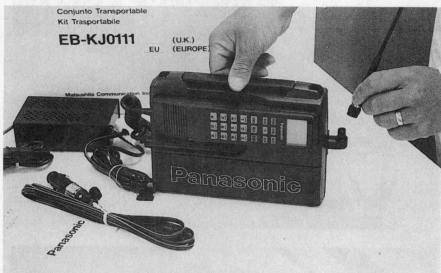

On now to the fitting procedure. Chris has spent many years fitting car phones and knows that one of the most common mistakes made by amateurs, including some people who make their living this way, is to drill through the petrol tank when attempting to mount the transceiver. In this case, we didn't go anywhere near it, but a professional needs to know exactly where it is.

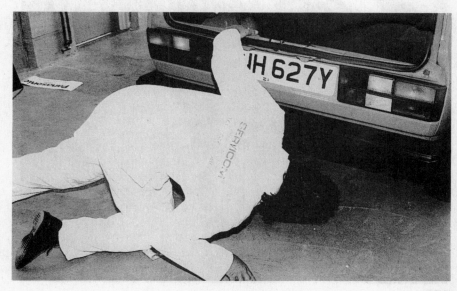

155

In-car telephones

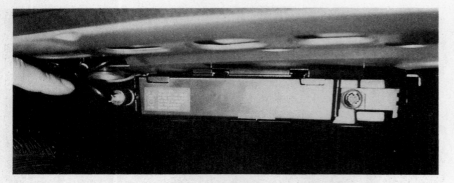

Chris always prefers to put the transceiver at the back of the car (hatch or boot) as it is generally safer there, and not likely to be kicked or knocked. In a saloon, there is often space to mount the transceiver under the rear shelf. This is ideal from all points of view.

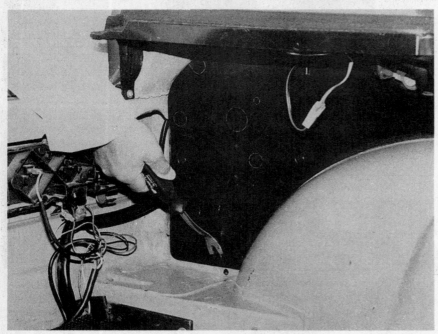

However, these options may not be open to you, either by design or lack of room, and you may have to opt from under-seat mounting. If so, remember to check for under-floor brake lines and cables before you drill any holes, and don't forget to use plenty of anti-rusting agent. Like hi-fi power amplifiers, the wiring should be pointing away from the rear passengers feet. In the case of this hatchback, the choice is floor mounting or, much better, on the side of the vehicle. Chris started by pulling back the rear trim panel to see what was behind it. The screwdriver-like item Chris is using is a Sykes-Pickavant trim-removing tool, a most useful piece of kit, saving broken fingernails and broken trim fixing studs alike.

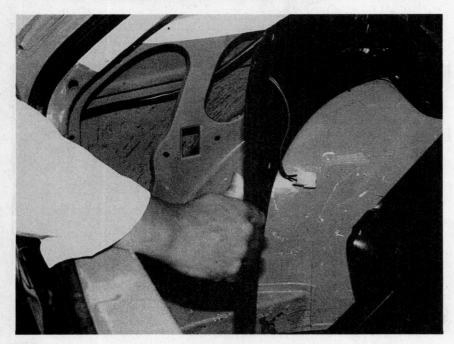

He has to check that there is no wiring in the way and that there is some metal behind the trim so that the transceiver can be mounted securely. He noted the single loom running along the top, but it was happily out of the way for our purposes.

In-car telephones

The careful measuring and marking procedure was followed by the drilling of four holes which—

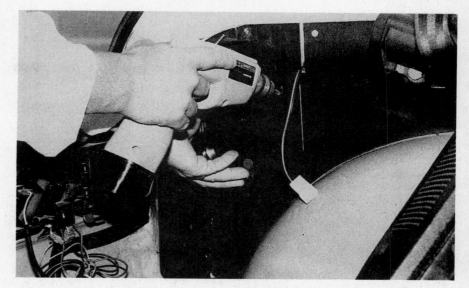

—facilitated the fitting of the transceiver bracket by means of self-tapping screws. Chris made sure that the transceiver would fit easily into the bracket and not foul the rear wheel arch before he started drilling!

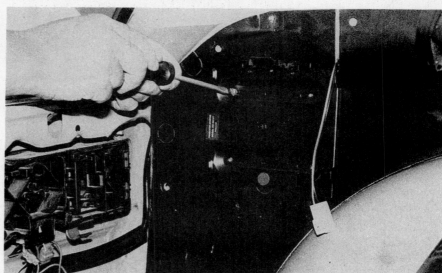

Then, it's onward and upward - to the roof and the aerial mounting. Yet more careful measuring is required. Measure twice, drill once - a good adage to remember at this point!

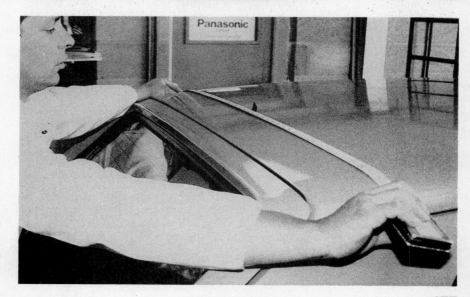

In-car telephones

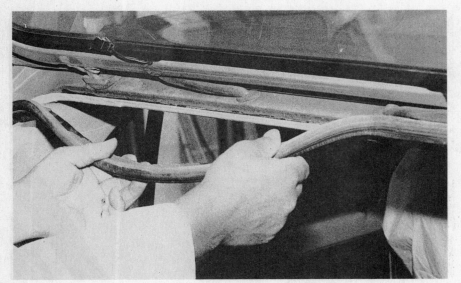

It's no use drilling a hole in the roof with the headlining still in place. Pulling down the rear headlining without ripping and, moreover, getting it back so you can't tell the difference, is the mark of a professional. The mastic sealant used along the join made this a very time-consuming task, to say the least.

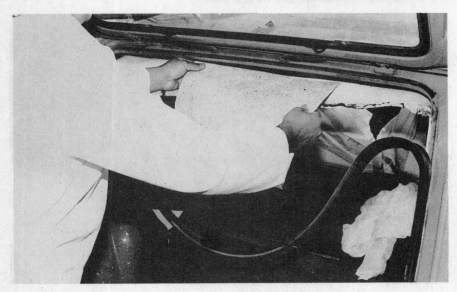

The hole is still not ready to be drilled. Chris slid a piece of sheeting between the headlining and roof in order to catch the swarf and provide a 'safety net' in case the drill slipped and came through too far.

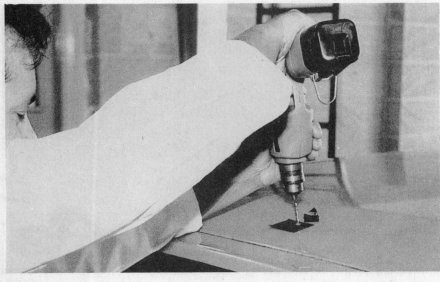

Note the use of tape to prevent the drill bit skidding across the paintwork whilst the hole is being drilled. After brushing the swarf away—

In-car telephones

—a small wire brush was fitted to the cordless drill and used to clean up the underside of the hole to ensure that there was a good, paint/grease-free earth for the aerial.

Like your standard radio tuner, the apparatus is only as good as the aerial that feeds it the signal. Allgon is one of the most respected names in the business and these are just two of their product range. The smaller, 1/4 wave 'golf ball' type on the right looks trendy and works very well in a city and/or built-up traffic environment. For better overall coverage and motorway work, go for a high-gain, co-linear aerial, like the one on the left which was fitted here.

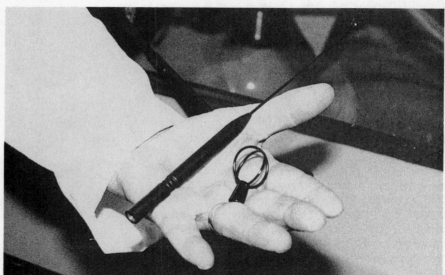

For those who have a fully mobile phone, there is much better reception to be gained by using a clip-on window aerial like this one. Alternatively—

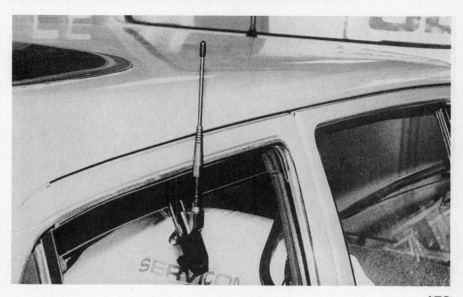

In-car telephones

Our roof mounted aerial was fitted in much the same way as a conventional FM aerial, with the same amount of care being taken not to over-tighten at this point; you could easily damage the aerial and the roof as well.

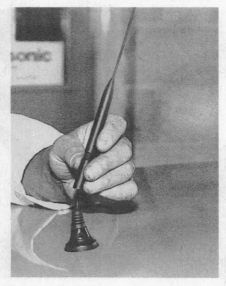

—-you could use a magnetic 'mag-mount' aerial and place it in the centre of the roof. Not the most elegant solution, but it does put the aerial in the right place and it offers much better reception for the portable hand phone.

The aerial itself screws into place and, therefore, unscrews for security purposes and/or visits to the car wash. It is also flexible at its base so that it can be bent to the best possible angle. This is always 90 degrees to the car; it may look good to have an aerial swept back in the current aerodynamic fashion, but the right angle for the best reception, is a right angle!

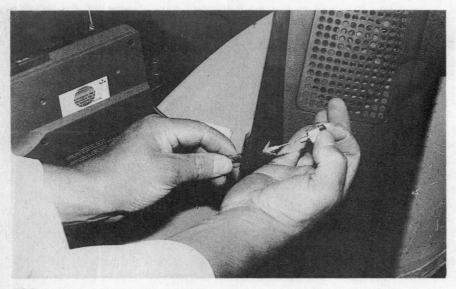

Having routed the cable down from the aerial, beneath the headlining and behind the 'C' pillar, it has to be cut to length and the coaxial plug fitted. The opportunities for messing up the whole installationhere are rife. It is so easy to short out the wires during the fitting of this plug.

160

In-car telephones

The wiring has to be taken from the rear to the front of the car. Chris placed electrician's tape at 6" intervals to make a loom of the wiring and used plastic cable ties to attach it to the vehicle loom. Don't forget the basic rules with regard to running power carrying leads alongside speaker wires. Note the way that Chris has pulled the carpeting right back so that he could see exactly what he was doing. This technique was repeated all the way to—

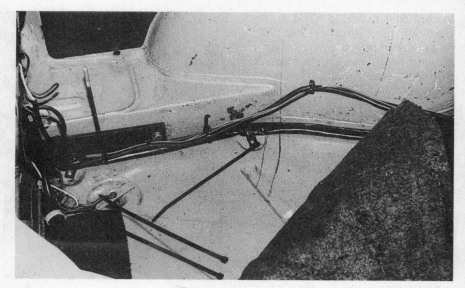

—the passenger side footwell, where the phone wiring was divided, with some wiring staying inside the car and the earth and power leads taken into the engine bay. In this case, Chris used an existing hole in the bulkhead. It was already grommeted (an essential point) and there was room to include a couple of extra wires. If you do have to drill a hole, make sure that you apply some anti-rust agent as well as that grommet.

The wiring was routed around the edge of the engine bay, following the vehicle main loom. The earth lead was connected before the power lead (both directly to the battery) and Chris made sure that the transceiver was disconnected before powering up. Note the in-line fuse.

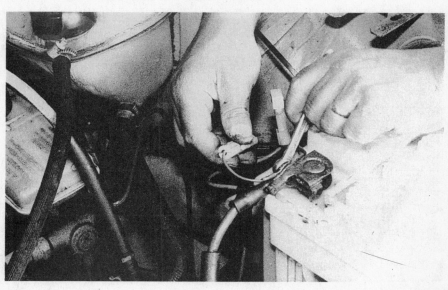

In-car telephones

Where to mount the handset cradle is a common problem. It has to be where the driver can reach it and where he can see it, but not where it will actually obscure his vision or interfere with his control of the car. There are few ideal places and, almost without question, compromises have to be made. One point to consider is the effect of drilling mounting holes in your dashboard. A dashboard full of unsightly holes is hardly a selling point and, even if you leave the phone installed for the next owner, he may not want it. The 'holey' problem remains! In this case, Chris decided to mount the cradle on the side of the after-market centre console. It's a bit of a stretch for the driver to reach it. No problem for the passenger, though and the driver should never use the handset whilst the vehicle is in motion anyway; there is no call important enough to warrant causing an accident through inattention to the job in hand - driving!

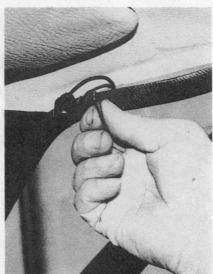

The wiring to the handset was routed underneath the dash and the only connection required is one to an ignition fed 'live' power source. The wiring from the hands-free microphone was routed beneath the driver's side dash and then up the 'A' pillar, under the rubber sealing strip. If you do this carefully, you'll find that this sealing will push back into its original position. Be careful, though, not to damage the phone wiring on the sharp edges of the body pressings. The microphone itself can be mounted here or on the sun visor. The quality from the tiny Panasonic 'mike' has to be heard to be believed.

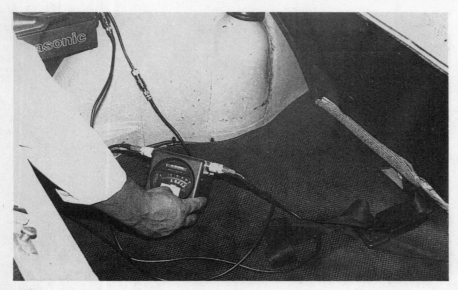

A Standing Wave Ratio (SWR) check has to be made. Here, the SWR meter is wired between the aerial and the transceiver. The reading will show whether all the connections have been made correctly especially the aerial plug). Even if you fit your own phone, you are well advised to pop down to your local fitting centre for them to run this simple test before you link it all up. You can easily damage its delicate innards by not doing so.

In-car telephones

The finished result, with the phone on its cradle and the trim replaced. All-in-all, a four hour job, done properly by a professional; and it shows. I would emphasise again, if you are not fully competent, do not start the fitment.

The basic wiring diagram shows that the electrical side is relatively straight forward. However, as we've already seen, there's more to installation than just wielding a few Scotchloks!

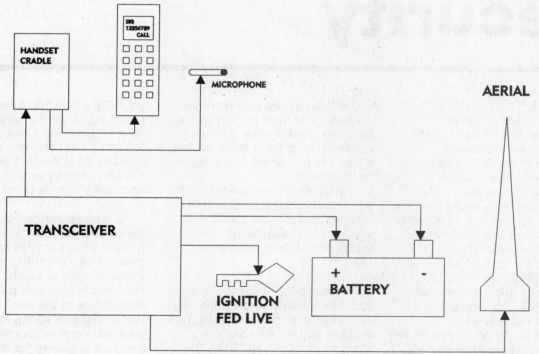

WIRING DIAGRAM: PANASONIC 'E' SERIES PHONE.

163

Chapter 13
Security

Car theft statistics make horrifying reading (unless you're a car thief!) and are the reason that an increasing number of manufacturers are fitting alarms as standard equipment to new cars. The insurance industry looks on cars equipped with alarms in a better light. Consider the following;

Thefts from and of cars accounted for 25% of all crime and cost ú390 million in 1991.

490,000 cars were reported stolen in 1990, of which 1 in 4 was never recovered.

A stolen car is 200 times more likely to be involved in an accident.

Of all British owners, 15% have experienced theft from their vehicle.

On average, a car is stolen every two minutes, day and night in the UK. It can take as little as 4 seconds to break into the average car.

Even with manufacturers becoming increasingly aware, most cars can be broken into in less than 30 seconds.

Car thieves are opportunists. Owners of high performance cars are, on average, 3 times more likely to have them stolen, some are ten times more likely to be stolen!

In-built security devices such as central locking and double locking and alarms, can significantly reduce the incidence of theft from and of vehicles.

The majority of car thefts tend to be concentrated on relatively few model ranges, your insurers will tell you which cars are classed as 'high risk' from the theft point of view.

All these statistics may lead you to trade-in your car altogether and buy a bus pass! However, it's not quite as bad as that and owners can do plenty to ensure that their cars and contents are where they left them. A massive percentage of reported crimes are theft **from** cars rather than theft **of** cars. Any guess which items of equipment are first in line? Yes, your ICE equipment makes a tempting target for anyone with light fingers and a need to find tonight's beer money; what you need to do is make it sufficiently difficult or risky to take your set and make him pass on to the next car.

As well as various security items you can buy, you can protect your car with something that is free - common sense! Park your car sensibly, remove removable sets and put CD/tape collections in the boot or at least out of sight. Parking in the darkest, loneliest corner of the car park is asking for trouble - you might as well leave the doors unlocked to save them the trouble of breaking a window!

What a thief does not like is risk and hard work - if you park your car under a street lamp, where plenty of people pass by, you make it risky for him to try and enter your vehicle. If you've removed the head unit from the dash, then it may be under the seat or in the boot; but equally,

Security

you may have taken it with you - more risk. The next step is to make your car too much like hard work to bother with. Without a doubt, a good, well-fitted car alarm is an excellent way of discouraging a potential thief. People who steal ICE equipment from cars are opportunist and, by definition, will not usually have the time to mess around. So, whilst it is true that few security measures are likely to be perfect, if you can make it difficult for him, he may just not bother

Don't think, just because your car has a fairly basic system, that the thief won't bother. Most stolen ICE equipment changes hands very cheaply indeed and most thieves will be happy to make a tenner on the deal. That deal could cost you hundreds in repair costs and/or insurance excesses and loss of no-claims discount.

An unalarmed car, with a non-coded radio/cassette deck grinning happily in the DIN aperture, is a sitting duck. It's up to you to make sure your car isn't on the target range.

The manufacturers themselves are well aware that radio/ cassette decks and/or ancillary equipment are high on the priority list of any opportunist thief and, over the last few years, have taken strong measures to prevent them doing so.

Coded decks were introduced. A personal code (usually 4 digits) is entered by the owner when the deck is installed. After this, should power be disconnected the same code has to be entered before the deck is operable. With a typical four-number code, there are around 10,000 different combinations to try. This security method has since been reinforced as the manufacturers have uprated the circuitry to prevent electrical hackers from getting inside and 'breaking' the code. According to Philips, the myth that popping a deck in the freezer overnight to break a code is just that; a myth!

WHAT TO LOOK FOR IN AN ALARM SYSTEM

Get an alarm with as many features as you can afford. Using pin switches to protect the doors is the first step - but many thieves break a window to reach in and steal things. So ultrasonic or microwave sensors are highly recommended, either of which will trigger the alarm should they detect an unwanted presence inside the car.

Microwave sensors are of particular use with convertible vehicles, but you must make sure that they are legal - some aren't. A shock sensor is useful as is a tilt sensor which triggers the alarm if the angle of the car varies by more than 2 or 3 degrees. The latter is designed to prevent the more organised thieves simply loading your car onto a trailer and driving off.

There should be some form of visual indication of the presence of an alarm. Window stickers should always be used and most alarms come with a flashing LED light which confirms that the system is fitted and working. With some alarms, removing a battery lead, or even just cutting the main power feed, kills the alarm. This is why you should prefer a system with battery backup, whereby removing the source of power actually triggers the alarm and leaves all its functions operational. Prefer a remote control system. A radio control alarm will have a better range, but must conform to licence exception 1340. As a rule, each handset will carry a sticker saying just that; if it doesn't then it's probably illegal! Some remote controls use infra-red, like your TV or video. These have a much shorter range and can sometimes be reflected off the glass in bright sunlight. Can your alarm be upgraded? Some offer the facility to add interfaces (electronic links) to switch electric windows, central locking and/or an electric sunroof.

Some form of vehicle immobilisation is a good idea, for there are cases where a brazen thief has actually driven a car away with the alarm sounding!

From a safety point of view, it is vital that the alarm you fit is wired so that it cannot be switched on whilst the vehicle is in motion.

INSURANCE

It is foolish to assume that your stereo system is included in your car insurance policy. Some policies include the *standard* system, but some include no hi-fi equipment unless you specifically ask them to. If you have problems getting your equipment insured, some household policies can be extended to include car head units. Whatever, you need to know what is and what isn't covered, before your car is vandalised!

The simplest security measure is to take the deck with you, and almost all equipment of any value is removable or 'quick-out' as many companies call it. Carrying the set around is made easier by using a Caselogic padded case. It pays to concentrate, though; there are numerous cases of sets being taken, say, into a restaurant and left there after the meal.

Security

Recently, the removable set principle has been taken a step further with the introduction of sets where just part of the set was removed. This Clarion radio/cassette deck has a removable front panel which fits neatly into a purpose-made pouch. The company also produce the ICE 'safe'.

Blaupunkt are so confident in their Keycard coding system, that they will provide a replacement deck if you buy one and have it stolen.

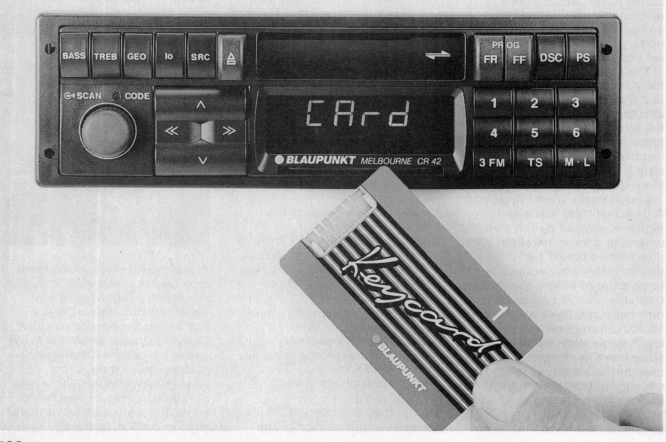

Security

Roadstar have gone a step further with the detachable front idea; with the front removed, an LED warns the thief he's in for trouble. If he ignores this and removes the deck from the car, an in-built siren sounds and keeps sounding for 30 minutes. Running down the high street inconspicuously with a 100 dB siren sounding does not make for the easiest of getaways! Because the siren is not actually part of an alarm system, it is exempt from the ruling stating that systems have to reset within 30 secs.

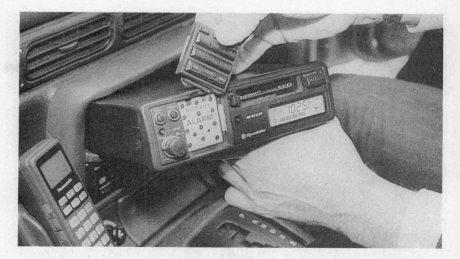

One way to protect a deck which is not removable, in whole or in part, is simply...

... to cover it up with a DIN size plastic cover. It is not possible to see what, if anything, is beneath. It could just as easily be a top-range deck as a £40 'special'. It introduces an element of risk into the car thief's life, which is what he doesn't want.

Security

For use with non-removable decks, or removable decks with lazy owners, there are number of locks designed to keep cassette decks in situ, of which three are seen here.

The Audiolok from Quickfit 70 requires 10 minutes or so of your time to fit the special cage into your DIN aperture. When the front cover is locked, four locking bars (one in each corner) combine to make it a head unit 'vault'. Like the Metro device below, it uses a radial pin tumbler lock.

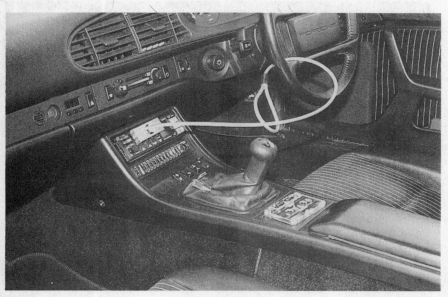

The Metro radio/cassette lock is a cassette shaped item attached to a very strong cable. With the 'cassette' looped through the cable and around the steering wheel, as here, it can be locked into the cassette deck. Removing it means virtually destroying the set and the cable is designed to resist bolt-cutters etc. It can be used with permanently fixed decks as well as removable ones where it is not convenient for the owner to take it away.

Security

The Motus security lock is another cassette shaped lock. It slots into the deck aperture and, on the turn of the special key, two clamps lock over the capstans. A large white 'tongue' with the word "ALARM" sticks out from the deck to warn the thief that there are no easy pickings here. As with any form of security...

... it is important to let the thief know that there is nothing to be had before he breaks the window to find out for himself. When you get a window sticker, use it!

The Sonix alarm is something of a hybrid unit. Essentially, a pair of ultrasonic sensors and a high power siren are housed in the main unit which clamps onto the steering wheel (any rim size up to 31mm). It takes a trickle charge from the cigar lighter socket and if power is disconnected, the alarm sounds just like a normal battery backup system. The ultrasonics also serve their usual purpose, detecting movement inside the car. It protects the inside of the car very well and has the advantage of being totally portable and so can be used in another vehicle or even a caravan or boat.

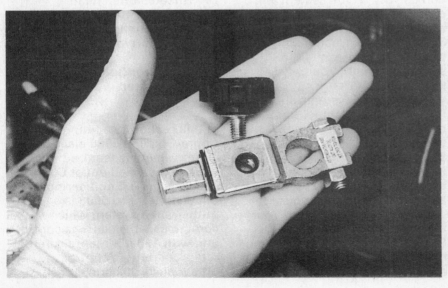

The Richbrook Dis-car-nect has a self-explanatory name. It is a device which sits between the earth terminal on the battery and the main earth lead. By removing the plastic knob, the battery power is removed. As such, it is a simple security device and handy, too, for when you are fitting your ICE equipment to save you messing around with spanner to take off the earth terminal every time.

169

Security

Alarm systems come in many formats, from the stunningly complex to the disarmingly (sic) simple. This Patrol Line model is a well-made unit which should be relatively easy to fit. However, the same caveat applies; it must be fitted correctly otherwise it will just be a nuisance.

FITTING A FULL ALARM SYSTEM

Fitting a full alarm system is highly recommended, even if you have a car that is not particularly valuable.

Just look at the excess on your insurance policy and then calculate how much no-claims discount you'd lose over the next two years by making a claim for a stolen or damaged vehicle. In many respects, a good quality alarm is an insurance policy for your insurance policy! Note, there, the word 'quality'; it is required in terms of both the product and the expertise of fitting. The alarms that false-alarm in the middle of the night are invariably poor quality alarms and/or units which have been fitted or set-up badly.

Fitting your own alarm is possible, given that you are justly confident in your ability to work to a high standard. Time is very important and lack of it is a key element in the poor fitting of many DIY alarms. Remember, for a full system, you will have to remove a surprising amount of trim and spend more than a little while looming and routeing the wires.

For making electrical joins, soldering should be preferred. If you are not proficient, then securely crimped joins are the next favourite. It is not recommended that you use Scotchlok joins in an alarm installation, especially in the engine bay, where the ingress of moisture could soon cause problems.

The 'before you start' checklist

1. Make sure that the alarm in question is genuinely suitable for your car. Ask the dealer if it will fit your particular vehicle - if the answer is not a definite 'yes', then buy a different alarm.
2. Make sure that all the component parts are present and correct. A full system will contain plenty of wiring, accessories and terminals.
3. Read the instructions thoroughly and don't lift a screwdriver until you fully understand them!
4. Most alarm systems supply the terminals required, but it's not a bad idea to have a few spares.
5. 3 spare fuses, just in case.
6. You must have enough time to fit the alarm correctly. Allow at least 4 hours for a typical full system. If you are linking your alarm to other accessories (electric windows etc.) then you should look to investing a whole day. Don't forget that trim removal and the neat looming of wiring takes an inordinate amount of time. You should allow for some fault-finding and, more importantly, correct setting-up. A rush job will more than likely show itself at 2 o'clock the following morning!
7. Unless you have a large garage, you should choose a fine day for alarm fitting, as you will find it more than useful to have the doors wide open.
8. Tools for the job. A typical alarm fitment will require;
Electric drill/bits
Rust proofing agent
Crimpers/terminals
Masking/electrician's tape
Crosshead/slotted head screwdrivers
Various spanners/socket set
Tape measure/pencil/scriber
Cable ties
Soldering iron/solder
Electrical tester
Pliers

In this section, the fitting of a Cobra Goldline 5906 electronic alarm system is featured. The work was carried out at Cobra specialist, GTi Engineering of Silverstone. The 5906 is a top-of-the-range system which comprises; an electronic control unit with 120 dB integral siren, radio key remote control, ultrasonic protection, shock

Security

sensor and the various accessories such as self-tapping screws, spare keys etc. required to fit it and use it. Twin ultrasonic sensors are part of the package. One sensor emits an ultrasonic signal and the other receives it, after it has passed around the car. If the ultrasonic beam is broken when the alarm is set (say, by a hand appearing through a window) the alarm will sound. It is important that the sensitivity is set-up correctly, otherwise false-alarming will be the order of the day. If you have a convertible car, then micro-wave sensors which are not sensitive to wind pressure, should be used, so that the alarm can still be set with the roof down.

It has an integral interface to control central locking and separate interfaces can be added to link it to an electric sunroof and/or electric windows. The totally automatic alarm, where all those options are incorporated, is a definite aid to security; at the press of a button your car becomes totally secure. Windows, doors and sunroof are locked and the alarm set and with a further press, the doors unlock and the alarm is switched off. It's so easy, there's no excuse not to do it every time you leave the car, however short the period. That way, you can be fairly sure it will be there when you return.

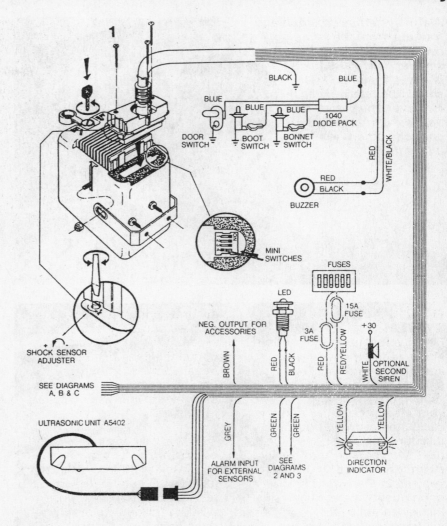

... schematic wiring diagram. It's not difficult if you're prepared to work steadily and do the job right. (Courtesy Italaudio Ltd)

It's a good idea to familiarise yourself with the component parts first. Couple this with some time spent studying the...

Security

This is the back of the combined siren/control unit, showing the multi-pin socket. The corresponding plug is in the end of the huge loom provided with the kit.

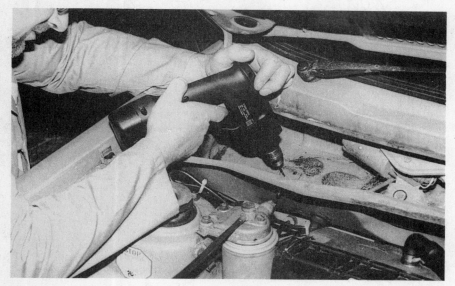

The siren/control unit has to be mounted out of the way of excess heat and moisture. In the front scuttle isn't a bad place and in many cases, there will be little choice. The positions of the two holes for the mounting bracket were marked and dot punched, then pilot holes drilled. The cordless drill made this a simple task.

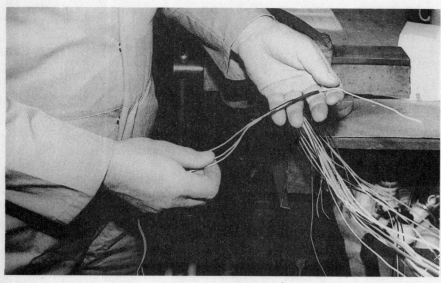

One of the first jobs is to straighten out the loom and tape it up as to used wires, unused wires, those staying in the engine bay and those which will be routed to the cabin. This is typical professional forethought you would do well to copy.

Security

The loom plug has to be inserted into the control unit before it is fitted because there are two securing screws to be tightened. This, combined with a water tight seal, ensures that there will be no problems caused by fittings working loose or the ingress of moisture. Note the solder in the background; as you would expect with a professional fitting, all joins were soldered.

In order to get the loom into the cabin area, another hole had to be drilled, a job for the cordless drill again, this time with the S/P Varicut drill. The newly cut hole was filed to remove the rough edges and then painted with rust proofing agent.

The control/siren sits neatly here, with the wiring loom laying out of harm's way to the right.

Security

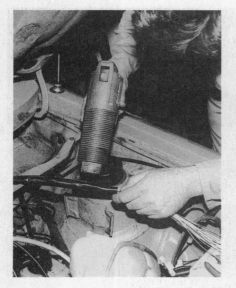

A hot air gun was used to shrink-wrap the looming tape around the multitude of wires. Apart from keeping moisture out, it also makes for a very tidy installation. Unfortunately, most DIY alarm fitments look like an explosion in a wire manufacturers!

All-important! Whenever you are passing wiring through a hole in the bodywork, whether it's a large loom or a single wire, you must use a rubber grommet, otherwise the plastic wire sheathing could be chafed through, leading to the possibility of short circuits.

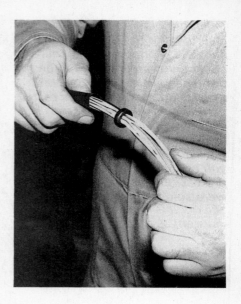

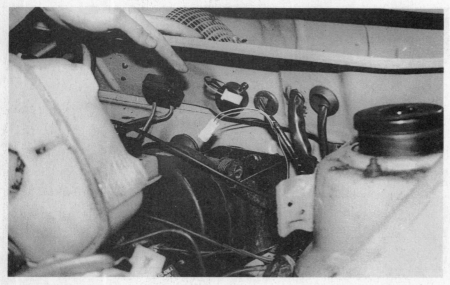

The Cobra system has an option of a confirmation bleep when the alarm is switched on and off. It comes not from the main siren, but this tiny bleeper, mounted here, above the brake servo.

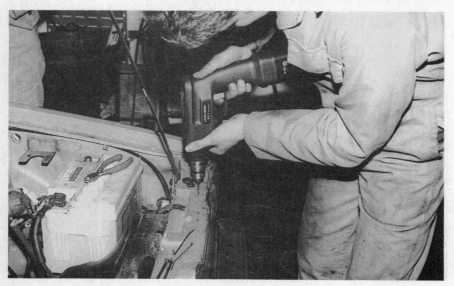

A bonnet pin switch has to be fitted so that the alarm sounds if the bonnet is lifted, which means measuring the precise distance on the edge of the bonnet and exactly the same distance on the slam panel. Only then can a suitable hole be drilled. Depending on what is fitted to your car as standard, you may have to fit pin switches to the doors and boot/hatch, or you may simply be able to wire into the existing ones.

Security

When the loom was brought into the cabin from the engine bay, it became apparent just how many pieces of wire there were! This model has a built-in interface for central locking. The car did not have it as standard, but an after-market kit was on the list of 'things to be done' so the wiring was coiled and taped out of the way behind the centre console. That way, it was ready and waiting to be linked to the auto-locking system at a later date.

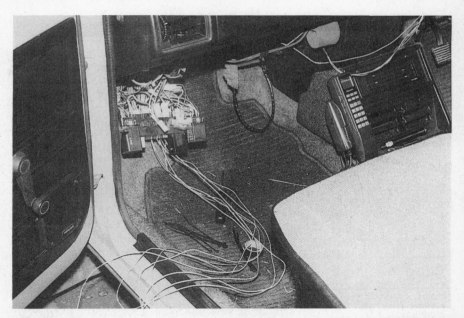

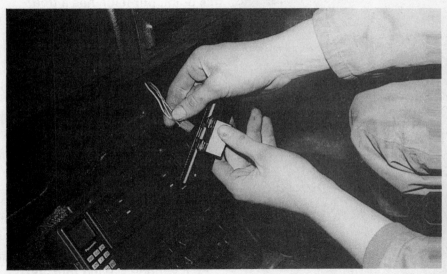

The ultrasonic sensors came in a single piece unit, light enough to be stuck into position. In this case the wiring was routed neatly behind the lower dash and the unit positioned just above the centre console. The flashing LED warns that the system is armed. If the system is activated during your absence, the LED flashes in one of a number of patterns to show exactly which part of the system triggered it (ultrasonics, bonnet pin switch, shock sensor etc.).

The Cobra remote key has two buttons. Pressing the top button activates the alarm, whereas pressing the lower button within a further 20 seconds de-activates the ultrasonic sensors, in-case you want, say, to leave your dog in the car, but have all the other alarm functions operational. On its own, the lower button becomes a panic button. It activates the siren instantly, regardless of whether or not the alarm is set. It serves a useful defence against muggers and the like, especially for the lady driver.

Security

Cobra provide two window stickers. It's in your interest to use them in order to stop a thief breaking your window before he realises there is an alarm fitted.

SECURITY TIPS

If your top-of-the-range alarm has been expertly fitted and you find it triggering time and again one evening, remember the 'clever' thief's trick; he rocks or knocks your car to set-off the alarm and then hides out of sight. You come out, switch it off and reset it. After few of these, you are tempted to switch it off altogether, usually in the cause of neighbourly relations and on the grounds that there must be a fault. At this point, our felon has a free hand - a car to work on which he knows is definitely not alarmed!

Don't leave valuables on display inside the car, lock them out of sight or take them with you.

Always lock your car, even if you only leave it for a couple of minutes - remember, most thieves are opportunist.

Always remove the ignition keys from your car - even when it is in your garage.

Be alert to a stranger taking interest in your vehicle.

At night park in well lit and frequented areas.

Etch windows, wing mirrors, lights and radio or hi-fi system with registration or other unique number.

Appendix: Useful Names and Addresses

Acoustic Research – See Jetcell Ltd

Aiwa, Unit 5, Heathrow Summit Centre, Skyport Drive, West Drayton, Middx., UB7 0LY. Tel: 081 897 7000 Fax: 081 564 9446
Full range of in-car audio equipment

Allsop CD/cassette care – see Path Group plc

Alpine Electronics of UK Ltd., 13 Tanners Drive, Blakelands, Milton Keynes, Bucks MK14 5BU. Tel: 0908 611556 Fax: 0908 618420
A wide range of in-car audio equipment

Altec Lansing – see Jetcell Ltd

Audioline – see Harry Moss International Ltd.

Auto Acoustics, Unit 18, Staines Central Trading Estate, Staines, Middx., TW18 4UX. Tel: 0784 460030 Fax: 0932 336103
A range of custom-made acoustic rear shelves and door,trim modules,

Auto Europe Ltd., Unit 6, Minden Road, Kimpton Road Trading Estate, Sutton, Surrey SM3 9PF. Tel: 081 641 1999 Fax: 081 641 5470
Keystop alarm systems

Autoblock security – see Tritec

Autocar Electrical Equipment Ltd., 77/85 Newington Causeway, London, SE1 6BJ. Tel: 071 403 5959 Fax: 071 378 1270
Meta alarms, Hirschman aerials, Becker car audio

Autoleads Limited, 11B Haven Way, Farnham, Surrey., GU9 9QU. Tel: 0252 735662 Fax: 0252 735813
Monitor power and speaker cable, plugs and distribution blocks.

Useful names and addresses

Automaxi Ltd., Chiltern Trading Estate, Grovebury Road, Leighton Buzz, Beds LU7 8TU. Tel: 0525 383131 Fax: 0525 370443
Motus cassette security lock

Bazooka Tube – see Path Group plc

Becker – see Autocar Electrical Equipment Ltd

Blaupunkt – see Bosch, Robert Ltd.

Bosch, Robert Ltd., PO Box 98, Broadwater Park, Denham, Middx., UB9 5HJ. Tel: 0895 838383 Fax: 0895 838388
Bosch security products, Blaupunkt in-car entertainment

Busybody Products, Unit 5, Maple Industrial Estate, Maple Way, Feltham, Middx., TW13 7AW. Tel: 081 753 1499 Fax: 081 890 2941
Suppliers of 'Lighthouse' portable lamp,

C.R.I.S.P., Communication House, 1 Kings Road, Crowthorne, Berks., RG11 7BG. Tel: 0344 761272 Fax: 0344 761292
Car Radio Industry Specialists Association

Car Hi-Fi Magazine The Evro Publishing Co Ltd., 60 Waldegrave Road, Teddington, Middx., TW11 8LG.
Tel: 081 943 5943, Fax: 081 943 5871
Bi-monthly in-car audio magazine

Car Stereo & Security Magazine, DIR Publishing Ltd., PO Box 771, Buckingham, Bucks MK18 4HH. Tel: 0280 812197, Fax: 0280 815633
Bi-monthly in-car entertainment, security and communication magazine

Caselogic – see Path Group plc

CEL Sales Ltd., CEL House, Unit 2, Block 6, Shenstone Trading Estate, Bromsgrove Rd, Halesowen, W. Midlands B63 3XB.
Tel: 021 585 6505 Fax: 021 585 5657
Manufacturers of CEL alarm systems

Clarion Shoji (UK) Unit 1, Marshall Road, Hillmead, Swindon, Wilts, SN5 7DW. Tel: 0793 870400 Fax: 0793 875747
Wide range of in-car entertainment equipment

Clifford Electronics, 21 Abbeville Mews, 83 Clapham Park Road, London, WS4 7BX. Tel: 071 498 0200 Fax: 071 498 8808
Clifford vehicle security systems

Cobra Security systems – see Ital Audio

Connaught Interior Designs, Unit 1, Warnford Industrial Estate, Hayes, Middx., UB3 1BQ. Tel: 081 573 0997 Fax: 081 569 2542
High quality installation & interior trimming specialists

Coraltronic, Unit 1, Chalklin Business Park, Longfield Road, Tonbridge Wells, Kent, TN2 3UG. Tel: 0892 515020 Fax: 0892 511056
Interconti car audio products & accessories

Crimeguard Car Security Systems, 17 Pegasus Way, Bowerhill Industrial Estate, Melksham, Wilts SN12 6TR. Tel: 0225 790730, Fax: 0225 790724
Alarm systems

Dards Electronics Ltd., Erica Road, Stacey Bushes, Milton Keynes, Bucks, MK12 6HS. Tel: 0908 319900, Fax: 0908 313411
High quality car audio and security installers

Demon Products Ltd., Industrial Unit, 266 Selsdon Road, South Croydon, Surrey, CR2 7AA. Tel: 081 667 1300 Fax: 081 686 4077
Demon Alarm systems

Useful names and addresses

Denon – see Hayden Laboratories

Digital Vehicle Security Systems, Unit 1B, Saxeway Business Centre, Chartridge Lane, Chesham, Bucks., HP5 2SH. Tel: 0494 792499 Fax: 0494 778790
Vehicle security systems

Draper Tools Ltd., Hursley Road, Chandlersford, Eastleigh, Hants., SO5 5YF. Tel: 0703 266355
Wide range of quality tools

Drivewires – see Hayden Laboratories,

EDA Sparkrite Ltd., Beacon Buildings, Leighswood Road, Aldridge, W. Midlands. Tel: 0922 743676 Fax: 0922 743616
Vehicle security systems

Electrosystems, 19 Fairways, New River Trading Estate, Cheshunt, EN8 0NL. Tel: 0992 34428 Fax: 0992 38716
Spyball security systems

Fischer, Artur (UK) Ltd., Hithercroft Road, Wallingford, Oxon., OX10 9AT. Tel: 0491 33000 Fax: 0491 33286
Fischer C-Box tailor-made and universal in-car cassette,and CD carriers

Foxguard Car Alarms, 1 Sedgemount Industrial Park, Bristol Road, Bridgewater, Somerset, TA6 4AR. Tel: 0278 428473 Fax: 0278 427212
Car security systems

FSM Ltd., PO Box 54, Waterlooville, Hants., PO7 8TL. Tel: 0705 240287 Fax: 0705 269155
Mac Audio ICE equipment

Gelhard (UK) Ltd. Unit 1B, Demuth Way, Junction 2 Industrial Estate, Oldbury, Warley, W. Midlands., B69 4LT. Tel: 021 554 0664 Fax: 021 552 9112
Wide range of in-car entertainment equipment

Gemini Elettronica, 12 North Street, Droitwich, Worcs., WR9 8JB. Tel: 0905 794565 Fax: 0905 774861
Gemini alarm systems, Spal electric windows, Zendar aerials

Genexxa., – see InterTAN ltd.

Goodmans Loudspeakers, 1-3 The Ridegway, Havant, Hants., PO9 1JS. Tel: 0705 492777 Fax: 0705 470875
A range of in-car entertainment equipment

GT Auto Alarms (UK) Ltd. 16 Clinton Avenue, Hampton Magna, Warwicks., CV35 8TX. Tel: 0926 490959 Fax: 0926 495641
GT security systems

Gunson Ltd., Pudding Mill Lane, Stratford, London, E15 2PJ. Tel: 081 555 7421 Fax: 081 534 7433
A range of electrical testing equipment

Harman (Audio) UK Ltd., Unit 1B, Mill Street, Slough, Berks., SL2 5DD. Tel: 0753 576911 Fax: 0753 535306
Harman Kardon amps, JBL speakers and Pyle tubes

Hayden Laboratories, Chiltern Hill, Chalfont St Peter, Bucks, SL9 9UG. Tel: 0753 888447 Fax: 0753 880109
Denon range of in-car audio equipment, Drivewires acessories, MB Quart speakers

Hi-fon, Unit 3, Waterway Enterprise Park, Trafford Wharf Road, Manchester, M17 1EY. Tel: 061 876 0363 Fax: 061 872 7157
Hi-fon UK manufactured sub-bass tubes

Useful names and addresses

Hifonics – see Jetcell Ltd

Hirschmann – see Autocar Electrical Ltd.

Infinity UK, Gamepath Ltd. 25 Heathfield, Stacey Bushes, Milton Keynes, Bucks., MK12 6HR. Tel: 0908 317707 Fax: 0908 322704
Infinity speakers

Interconti – see Coraltronic Ltd.

InterTAN Ltd., Tandy Centre, Leamore Lane, Walsall, W. Midlands, WS2 7PS. Tel: 0922 710000
Tandy, Memorex and Genexxa ICE equipment.

Ital Audio Ltd., K & K House, Station Approach, Rickmansworth Road, Watford, Herts., WD1 7LR. Tel: 0923 240525 Fax: 0923 220011
Cobra vehicle alarm systems

JBL – see Harman Audio (UK) Ltd

Jetcell, 1A Berens Road, London., NW10 5DX. Tel: 081 969 2514 Fax: 081 960 5965
Acoustic Research/Altec Lansing speakers. Streetwires acc, HiFonics amps

JVC UK, Priestley Way, Eldonwall Trading Estate, London. NW2 7BA. Tel: 081 450 3282 Fax: 081 784 0131
Range of in car audio equipment

Kamasa Tools Ltd., Lower Everlands Road, Hungerford, Berks., RG17 0DX. Tel: 0488 684545 Fax: 0480 684317
Wide range of tools

KEF Electronics Ltd. Tovil, Maidstone, Kent, ME15 6QP. Tel: 0622 672261
KEF speakers

Kenwood – see Trio Kenwood

Keystop alarm systems – see Auto Europe Ltd.

Kingsland Robart & Co. 7 Cross Hill, Shrewsbury, Shrops., SY1 1JH. Tel: 0743 360545 Fax: 0743 360726
Specialist insurance brokers for mobile electronics

Laserline Car Alarms, 5 Beeston Road, Stuart Road, Manor Park, Runcorn, Cheshire, WA7 1SG. Tel: 0928 580560 Fax: 0928 580562
Security systems

Lighthouse portable lamp – see Busybody products

Lucas Automotive Ltd, Unit 7/10 Mica Close, Amington Industrial Estate, Tamworth, Staffs., B77 4QH. Tel: 0827 53344 Fax: 0827 53456
Mantis alarms, Lucas security systems, Radiomobile in-car audio equipment

M.A. Distributors, Industrial House, Conway Street, Hove, E. Sussex., BN3 3LV. Tel: 0273 720129 Maystor Security products.

Mac Audio – see FSM Ltd.

Maystar – see M.A. Distributors Ltd

MB Quart speakers – see Hayden Laboratories.

Memorex – see InterTAN Ltd

Merlin Aerials Ltd., Venture House, Bone Lane, Newbury, Berks., RG14 5SH. Tel: 0635 30001 Fax: 0635 35406
Nippon Antenna car aerials

Meta Security – see Autocar Electrical Equipment Ltd

Useful names and addresses

Modulus Car Alarms, Unit 14, International Business Park, Charfleets Business Park, Canvey Island, Essex. Tel: 0268 690002, Fax: 0268 684718
Vehicle security systems

Monitor cables/connectors see Autoleads

Monogram Audio, 25 Richmond Wood Road, Bournemouth, Dorset., DH8 9DG. Tel: 0202 513448
MTX in-car speakers

Moss, Harry International Ltd., Block D, Washford Industrial Estate, Redditch, Worcs., B98 0EA. Tel: 0527 584854
Fax: 0527 000000
Audioline in-car entertainment equipment, Moss security

MTX Speakers – see Monogram Audio

Nakamichi – see Path Group plc,

Nippon Antenna – see Merlin Aerials

Orvell Electronics, 17 Lancaster Road, Uxbridge, Middx., UB8 1AP. Tel: 0895 70137 Fax: 0895 71734
Piranha alarms

Panasonic Consumer Electronics, Panasonic House, Willoughby Road, Bracknell, Berks., RG12 4FP. Tel: 0344 853200
Fax: 0344 861656
A wide range of in-car audio & cellular communications equipment

Path Group plc, Unit 15, Hayward Industrial Park, Tameside Drive, Castle Bromwich, Birmingham B35 7BT. Tel: 021 776 7616 Fax: 021 748 3838
Caselogic cassette and CD portable storage & Allsop Tape and CD cleaning systems

Path Group plc, Unit 2, Desborough Industrial Park, Desborough Park Road, High Wycombe, Bucks., HP12 3BG. Tel: 0494 459981
Fax: 0494 461209
Distributors for Bazooka, Rockford-Fosgate, Nakamichi Phoenix Gold

Philips Car Stereo, PLS House, Talisman Road, Bicester, Oxon, OX6 0JX. Tel: 0869 320333 Fax: 0869 365005
Wide range of in-car audio equipment

Phoenix Audio – see Sonicare Ltd

Phoenix Gold – see Path Group plc

Pioneer High Fidelity, 1-6 Field Way, Greenford, Middx., UB6 8UZ. Tel: 081 575 5757 Fax: 081 575 3111
Wide range of in-car audio equipment

Piranha Car alarms – see Orvell Electronics

Pro-Plus Sound Ltd., Edgeware, Middlesex. Tel: 081 959 1982
Fax: 081 959 8691
Pro-Plus speaker range

Pye International Ltd., 420/430 London Road, Croydon, London., CR9 3QR. Tel: 081 689 4444 Fax: 081 684 8857
Range of in-car audio equipment

Pyle sub-bass tubes – see Harman Audio (UK) Ltd

Quickfit 70 Ltd, Alma Works, Fearnhead Street, Bolton, Lancs., BL3 3PE. Tel: 0204 62381 Fax: 0204 61941
Range of custom and universal centre consoles and Audiolok, cassette lock,

Useful names and addresses

Radiomobile – see Lucas Automotive Ltd

Roadstar (UK) Ltd., Roadstar House, Tavistock Industrial Estate, Ruscombe Ln Twyford, Berks., RG10 9NJ. Tel: 0734 321032 Fax: 0734 343011
Wide range of in-car audio equipment

Rockford-Fosgate – see Path Group plc

Sanyo – see Sonicare Ltd

Servicom Ltd., Unit 9, Old Forge Business Centre, Little Forge Road, Park Farm North, Redditch, Worcs., B98 7SF. Tel: 0527 510800,
Cellular phone suppliers and installers

Sharp Electronics (UK) Ltd., Sharp House, Thorp Road, Newton Heath, Manchester M10 9BE. Tel: 061 205 2333 Fax: 061 205 7076
Wide range of in-car audio equipment

Simba Security Systems Ltd., Occupation Road, Walworth, London, SE17 3BE. Tel: 071 703 0485
Vehicle security systems

Sonar Electronic (UK) Ltd., 272A Brixton Hill, Streatham, London, SW2 1HP. Tel: 081 671 6818 Fax: 081 671 6215
Sonar security systems

Sonicare Ltd., Unit 1B, Demuth Way, Junction 2 Industrial Estate, Oldbury, W. Midlands B69 4LT. Tel: 021 552 9797 Fax: 021 552 9112
Suppliers of Phoenix Audio, Sanyo and Blue Shield alarms.

Sony (UK) Ltd., Sony House, South Street, Staines, Middx., TW18 4PF. Tel: 0784 467142 Fax: 0784 463278
Wide range of in-car audio equipment

Sound Systems – see Ital Audio Ltd

Spal electric windows – see Gemini Elletronica

Sparkrite – see EDA Sparkrite Ltd

Spyball Security Systems – see Electrosystems

Streetwires – see Jetcell Ltd

Sykes-Pickavant Ltd. Warwick Works, Kilnhouse Lane, Lytham St Annes, Lancs., FY8 3DU. Tel: 0253 72191 Fax: 0253 713076
Wide range of tools and diagnostic equipment

Tandy – see InterTAN Ltd.

Teng Tools Ltd., Unit 5, Flitwick Industrial Estate, Maulden Road, Flitwick, Beds. MK45 1UF. Tel: 0525 718080, Fax: 0525 718882
Wide range of quality tools

Trio Kenwood UK Ltd., Dwight Road, Watford, Herts., WD1 8EB. Tel: 0923 816444 Fax: 0923 819131
Wide range of in-car audio equipment

Tritec Distribution, Heathrow House, 7 Raglan Road, Reigate, Surrey, RH2 0DR. Tel: 0737 223314
Autoblock vehicle security systems

Truck King, 320 St Albans Road, Watford, Herts., WD2 5PO.

Waso Ltd., Unit 1, Oakwood Estate, Mode Wheel Road, Salford, M5 2DS. Tel: 061 736 0767,
Vehicle security systems

Glossary of Terms

AM
Amplitude Modulation, which is just one way of transmitting audio on a radio signal. Medium wave and long wave stations are transmitted on AM. (See also FM)

ARI
An abbreviation of 'Autofahrer Rundfunk Information' an RDS-style traffic information system used in W.Germany, Austria and Switzerland. Sets capable of picking up the ARI signals often have a button marked 'SDK'. Though an RDS equipped set can utilise ARI signals, the reverse is not true.

Aerial
The device which picks up the radio signals being broadcast and passes them to the tuner. Sometimes called an antenna.

Auto eject
A cassette feature. When the tape has finished playing one side, it ejects automatically. Usually the set (in the case of a combination unit) will then revert to radio mode.

Automatic muting
A circuit included in some radios which silences the hiss encountered between stations.

Automatic same network following system.
An RDS feature. See under RDS

Automatic sound level control
A system whereby the volume of the sound equipment is increased automatically as the noise in the car increases (road noise, engine noise, tyre noise, etc.). Used by Blaupunkt and others who all use a different designation.

Auto-reverse
One of the most useful tape features. When the cassette reaches the end of the tape, the direction is automatically reversed and the other side played. It allows continuous tape listening without distracting the driver's attention from the road in order to turn the cassette over.

Auto-scan
This is a radio function where the tuner scans the waveband to find a station, but only holds it for approximately ten seconds before travelling to the next one. This allows the driver to keep his attention on the road whilst also searching for the station of his choice. When this is found, pressing the Auto scan button a second time maintains the station.

Auto seek
A similar feature to Auto Scan except that once a station is located, the tuner stops altogether. If the station is not the one you wish to listen to, you have to press the Auto Seek button again.

Autostore
A Philips designation for a radio

Glossary of terms

function where the tuner finds the six (number can vary) strongest signals and stores them in its memory. These are accessed by the preset buttons and having fitted a new set, this is usually the quickest and easiest way to program stations in. Other manufacturers have their own versions of Autostore, for example, Best Station Memory.

Azimuth
This is the angle of the recorded track in relation to the tape head. A good azimuth setting is required for a good frequency response and anything different from a 90° (right angle) will result in a 'dull' sound.

Balance
The balance control is effectively a right/left fader. Included on stereo units as an integral feature. (See also Fader).

Bass
Sounds in the low frequency range (up to approximately 600 Hz). See also 'Woofer' and 'sub-woofer'.

Bridgeable amplifier
A stereo amplifier on which both outputs can be combined to make a monaural amp of approximately twice the power rating. Also refers to four channel amplifiers where the outputs can be combined to make a more powerful two channel (stereo) unit.

Capstan
The capstan is the tape driving spindle which presses on the pinch roller and transports the tape.

Clipping
A form of distortion caused by over amplification.

Class A, B
A method of defining the way that an amplifier performs. Class A amplifiers produce extremely high quality sound, whereas Class B amplifiers are more efficient in producing more output from any given current output. As you may imagine, a Class AB amplifier seeks to combine the best of both worlds with a compromise solution.

Closed box
Refers to a totally enclosed speaker enclosure. Sometimes referred to as an infinite baffle.

Co-axial (1)
Speakers which have two, rather than one, drive units in one enclosure. One of the most common is to have a tweeter and mid range speaker together. The main advantage is that better frequency response can be obtained without needing extra space for speaker mounting. (See also Dual Cone speakers).

Co-Axial (2)
As well as speakers, this phrase can also be used to refer to twin core cable used for aerials. This contains a plastic sheathed inner wire surrounded by a braided outer. Should **not** be used for any other purpose.

Code
A security measure for protecting your ICE head unit. When fitted, a code is entered by the owner. If the power supply is disconnected at any time, this code has to be re-entered before the set will work. There can be as many as 10,000 permutations for a four figure code and this, together with certain extra safeguards with regard to timing, makes the set worthless if stolen.

Combination unit
Any ICE unit, usually DIN sized which houses more than one item, for example a radio/cassette player or a CD/tuner.

CrO2
Abbreviation of Chromium Dioxide, which is applied as a coating to some type II cassettes. (See Cassettes).

Crosstalk
See 'Stereo separation'

Crossover unit
Sometimes called a crossover network, this box of tricks distributes different frequencies to different loudspeakers, treble to a tweeter, bass to a woofer etc. An 'active' crossover divides the sounds before amplification, sending each set of signals to a separate amplifier. Conversely, a passive amplifier divides already amplified signals. The former gives a much purer sound, but is, of necessity more expensive and demanding in terms of fitting skill and space.

DAT
An abbreviation of Digital Audio Tape. The tapes are similar in essence to normal cassette tapes, albeit physically much smaller, but the sound signals are digitally encoded and they must be played on a DAT player. See also DCC.

dB (Decibel)
The dB is a unit referring to the measurement of the relative amplitude of different sound levels.

DBX
A noise reduction system which reduces tape hiss on playback. See also Dolby.

DCC
An abbreviation for Philips' Digital Compact Cassette. On a suitable machine, the DCC will produce CD quality sounds and, unlike the DAT (see above), it is the same size as the original compact cassette and backwards compatible.

Glossary of terms

DIN (Deutscher Industrie Normenausschus)
DIN E is the German standard adapted by almost all manufacturers for the measurement of ICE apertures. Also, the term is used to describe the five or seven pin plugs sometimes used for interconnecting various pieces of ICE. Earlier cars may have ISO DIN apertures, which means that DIN sized sets will not fit exactly, although there are still some manufacturers that make them.

DSP
An abbreviation for Digital Signal Processor. Effectively, this is a glorified graphic equaliser which is capable of doing marvellous things with digitised information. 'Moving' the sound around the car is one, and removing the lead singer's voice, thus creating a karioke effect, is another.

DX
Also referred to as 'local'. Changes the sensitivity of the tuner so that it selects only local stations or seeks out those further away, depending on its setting.

Defeat
This is a function of some graphic equalisers. It 'defeats' the equalisation controls in use and provides a 'flat' sound for comparison.

De-magnetisation
This is the process of cleaning a cassette head 'magnetically' in order to reduce the degradation in sound which occurs as particles from the tape build up on the head.

Dolby
A system designed by Ray Dolby for reducing tape hiss on playback. There are (at present) two types: Dolby 'B' and Dolby 'C'. The latter provides better results but tapes must be recorded/played on a similarly endowed and accurately matched equipment.

Dual cone speaker
A speaker which has a smaller cone inside the main cone. However, it is not a co-axial, which has two **separate** speakers.

Dynamic range
In music, this is the range between the loudest and the softest musical passages. It is expressed in decibels and the wider the range, the better.

EQ
An abbreviation for 'Equalisation' and can be used as a measurement of cassette tape performance of the process which divides the signal into desired ranges (notably in the case of a graphic equaliser) reducing frequency or phase distortion for high quality sound making.

FM
An abbreviation for Frequency Modulation and is used of the broadcast of stereo signals.

Fader
A control fitted to some sets which balances the sound of a four speaker system from front to rear. Sometimes fitted as an integral part of a set but if not, a separate fader can be added. (See also balance control)

Fast forward/fast rewind
A cassette deck function where the tape is sent quickly forwards or backwards. Sometimes lockable, these controls will FF or RW the tape until manually stopped or until the tape reaches the end. Some sets will then eject the tape, others will automatically play the other side and some will revert to radio mode (in the case of a combination unit).

Flat response
Uniform amplification across the full frequency range, i.e. with neither treble nor bass frequencies boosted or cut. See also **defeat**.

Frequency response (FR)
This denotes the span between the lowest and highest notes (or frequencies) that can be handled by the unit in question. All items of ICE, including speakers will have a specific FR and it follows that all units in any system should be compatible.

Front-end unit
A piece (or pieces) of ICE equipment, used as the first stage of a system (a sound source), without which the ancillaries are useless, for example, a radio/cassette deck

Fully automatic tuning
An RDS feature. See **RDS**.

Graphic equaliser
The 'graphic' is basically a sophisticated tone control By dividing the audible frequency spectrum into multiple bands they can be adjusted individually and thus provide much more accurate control of the sound. Graphics can have between 3 and 11 bands. Some have an integral amplifier, which can be useful, particularly from the point of view of mounting space. The unit gets its name from the way the band controls (or LED lights) form a graphic indication of the sound. (see **Spectrum analyser**)

Heat sink
A method of getting rid of unwanted heat from an amplifier. In ICE terms it usually takes the form of large cooling fins on the outside of the casing. On particularly high power amplifiers, an internal fan may be used.

185

Glossary of terms

Hertz (Hz)
Hertz or its two letter abbreviation, Hz, is the measure of frequency in cycles per second. The lower the number, the lower the note and visa versa.

Hi-fi
A standard abbreviation for High Fidelity. In an audio sense it means reproduction of sound exactly the same as the original. There is, in fact, a technical specification which systems should meet in order to be 'real' Hi-fi, DIN 45.500. The term is more commonly (though often inaccurately) used to mean any uprated ICE system.

IC
An abbreviation of Integrated Circuit. This is basically a 'computer on a chip' and in most ICE equipment, there are several, controlling various functions.

IAC
An abbreviation of Interference Absorption Circuit, another Philips development. It is designed to absorb interference (for example, from the car's engine) whilst the tuner is in FM mode. Many other manufacturers use this system under licence.

Impedance
Used here for the resistance value of a loudspeaker. It is important that the impedance of a speaker is compatible with that of the amplifier, whether the amplifier is a separate unit or part of the front end unit. It is measured in ohms and whilst the speaker impedance can be greater than the amplifier impedance, it should NEVER be less.

Infinite baffle
See **closed box**.

Key-off eject
This function releases the tape from the transport mechanism if the tape deck is switched off with the tape still inside.

Key-off pause
A similar function to that above, except that when the power is turned off, the cassette remains in the set. The tape is released from the mechanism and thus no strain is placed on either deck or cassette.

LCD
Abbreviation for Liquid Crystal Display used to display the functions of some front-end units.

LED
Abbreviation for Light Emitting Diode used to display the functions of some front-end units.

Laser pick-up
This is the optical device which reads the digitally encoded signals on a compact disc. In most sets there are three pickups, to ensure that tracking is absolutely perfect.

Last station memory
With this function, the radio will automatically tune itself to the last station played before it was switched off.

Local
See **DX**.

Logic control
A term given to the microprocessor control of tape deck functions which permit the use of 'light touch' controls. LC is much favoured as it almost guarantees safe tape handling.

Loudness
A function designed to boost low frequencies at low volume settings in order to compensate for the inadequacies of the human ear.

Maximum power output
The absolute maximum power an amplifier can produce, albeit for a very short period of time. The difference between this and the RMS (continuous) power rating can be considerable. Reference both to the THD figure.

Mid-range speaker
A speaker designed to handle those frequencies in the mid-range. Sometimes known by its American name of 'squawker'.

Multi-channel system
A system where separate amplifiers are used to drive individual speaker, or pairs of speakers.

Multipath interference
Radio interference caused by the same radio signal being received from different directions. The audio equivalent of 'ghosting' on a TV set.

Music repeat
A tape/CD function which causes the same track (or series of tracks) to be repeated until the function is cancelled.

Network indication
An RDS feature. See **RDS**.

Noise free balanced cable
This is what is says; specialised high quality cable which automatically cancels out any unwanted noise that it picks up. Usually used in uprated systems.

Noise reduction
A method of electronically reducing cassette hiss on playback. There are various system available from different manufacturers.

Oversampling
A piece of CD terminology which effectively produces a cleaner sound from the machine. Originally, two times oversampling was deemed

Glossary of terms

enough, but this has risen to four, eight and even sixteen times oversampling on some of the top CD players.

Phasing
An odd sound effect produced by incorrect wiring of positive and negative speaker terminals (either at the speaker or the set). As one speaker cone is moving out the other(s) is moving out. This has the affect of diluting much of the bass sound.

PLL
Abbreviation for Phase Locked Loop, a radio feature which locks FM signals on station and cuts out most interference.

RDS
Abbreviation for Radio Data Signalling. This is a traffic information system, broadcasting digital signals from the BBC and the IBA on the FM waveband which will include, among others, network indication, auto same network following system, fully auto tuning and information about road and traffic conditions. It is localised which means that the driver will get information about the area he/she is driving in. A special tuner is required to received the information. Still fairly new to the UK although a similar system as been in operation in Germany for many years.

RMS
Abbreviation for Root Mean Square, a technical term which defines the way the power of an amplifier is measured.

Removable sets
As the name suggests, these are front-end units which are designed to fit into the car via a permanently-mounted mother case, allowing the set to be removed when the car is left unattended. Various manufacturers have their own designations for such sets.

There are variations on the theme, notably those where just the front part of the set is removed, still rendering the remaining equipment useless, but meaning that there is less equipment for the owner to carry around.

Resonance
Vibration of a surface of space caused by sound waves. This usually occurs when excessive bass is used and/or speakers and trim panels are not secure.

RTA
An abbreviation for Real Time Analyser, an item of professional ICE test equipment designed to measure the acoustic characteristics of a specific vehicle.

Sensitivity
This is a measure of how a tuner can handle weak signals usually given in microvolts, with a different reading for FM, MW and LW. For a high quality set, look for 2uV, 12uV and 20uV respectively. An average receiver would probably turn out figures of 5 uV, 50uV and 40uV.

Signal strength
This relates to the strength of the radio signal being received by the tuner.

Signal to noise ratio (S/N)
This is the ratio of music signal level to the noise level generated by a component, e.g an amplifier. The higher the figure quoted, the better. Look for a S/N or 60 dB as an average and 90 dB as a really good figure.

Spectrum analyser
This is a graphic equaliser function where the graphic display takes the form of LED lights which respond to the level of audio signal in each frequency.

Sub-woofer
A speaker which reproduces frequencies from the ultra low bass region, up to approximately 200 Hz. These are usually 'driven' by separate amplifiers from the rest of the audio system. Because of the nature of bass frequencies, (i.e. they are felt rather than heard) they can be mounted almost anywhere in the car and are usually powered up in mono (as opposed to stereo) mode.

SWR
An abbreviation for Standing Wave Ratio. An SWR meter must be used to tune the aerial of a CB radio/cellular telephone before use. The meter measures the signal reflected back down the aerial cable during transmission. A high SWR could damage the transceiver.

Stereo separation
A figure, denoted in decibels, relating to the amount of isolation between the two stereo channels. It is sometimes referred to as crosstalk. The higher the figure the better and CD players will usually show a much higher figure than a radio or cassette player.

Tape scan
This is similar to the radio 'auto scan' feature. When activated, the tape will fast forward to the next track and then play approximately ten seconds of it. After that, it will repeat the operation until you decide to listen to a particular track, when pressing the function button again will switch it off.

THD
An abbreviation for Total Harmonic Distortion, a measure of the extent to which audio signals are distorted by the various pieces of electrical equipment as they pass through. It is given as a percentage figure, where 10% would be most noticeable and

Glossary of terms

1% audible to anyone listening for it. Extremely good (and usually extremely expensive) amplifiers can produce high power outputs at very low levels of THD, down to around 0.01%. When a power output is given, it should be allied to a THD figure, for example, 30W RMS at 0.1% THD.

Tone control
Any control used for cutting or boosting the treble or bass frequencies. In its simplest form it can be just one control turned one way for more treble and the other for more bass (by definition boosting one and cutting the other). Most sets split the functions and have one control for each. See also graphic equaliser.

Track search
This is a useful tape feature whereby when the button is pressed, it will fast forward to the next track. It will stop and play the tape from there on. It relies on being able to find the gap of silence in between tracks (it usually needs around 3 seconds) and is of little use where the gaps are small or non existent.

Traffic information reception
An RDS feature, see under **RDS**.

Transceiver
Similar to a tuner except that it can transmit as well as receive radio signals. Would be used in CB radios and in-car telephones.

Treble
Sounds in the high frequency range (approximately 4000 Hz upwards).

Tri-axial
Speakers which have three separate drive units in one enclosure (see also **co-axial speakers**).

Tuner
The part of the radio which allows different stations to be selected. Can be mechanical or electronic, although most modern sets are the latter.

Tweeter
A loudspeaker used solely for the reproduction of sounds in the treble range (approximately 4000 - 20000 Hz).

Watt
A measure of power output and consumption.

Wide band loudspeaker
A loudspeaker capable of reproducing the whole range of audio frequencies rather than a specific range (such as a tweeter or woofer). By definition, a wide-band speaker has to be something of a compromise.

Woofer
A speaker designed to handle low frequencies (up to approximately 600 Hz)

Wow and flutter
This is a term given to variations in cassette tape speed which can cause unpleasant howling sounds. Most cassette decks quote a figure and you need to look for a low one - the lower the better.

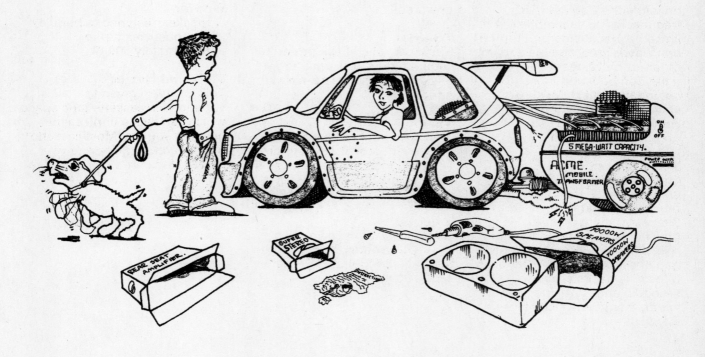

"Yes, I've just completed the system. Fancy a listen?" It's only a thought, but it is possible to get carried away ...